COSMOGENIC NUCLIDES
Principles, Concepts and Applications in the Earth Surface Sciences

This is the first book to provide a comprehensive and state-of-the-art introduction to the novel and fast-evolving topic of *in-situ* produced cosmogenic nuclides. It presents an accessible introduction to the theoretical foundations, with explanations of the relevant concepts, starting at a basic level and then building in sophistication. It incorporates, and draws on, methodological discussions and advances achieved within the international CRONUS (Cosmic-Ray Produced Nuclide Systematics) networks. Practical aspects, such as sampling, analytical methods and data interpretation are discussed in detail and an essential sampling checklist is provided. The full range of cosmogenic isotopes is covered and a wide spectrum of *in-situ* applications are then described and illustrated with specific and generic examples of exposure dating, burial dating, erosion and uplift rates, and process model verification.

Graduate students and practitioners will find this book a vital source of information on the background concepts and practical applications in geomorphology, geography, soil science and geology.

TIBOR J. DUNAI is a Reader in Geomorphology at the University of Edinburgh and has extensive research experience in the field of cosmogenic nuclides and their applications to Earth sciences. He was the initiator and coordinator of CRONUS-EU and is on the steering committee of CRONUS-Earth, and his methodological work on cosmogenic-nuclide production has been fundamental to the research of these networks. Dr Dunai has taught at several specialist training workshops, and organized the CRONUS-EU summer school on methodology and Earth science applications of cosmogenic nuclides. This book stems from these training events and is designed to address the expressed requirements of the participants, who included experienced practitioners as well as novices to the technique.

Cosmogenic Nuclides

Principles, Concepts and Applications in the Earth Surface Sciences

TIBOR J. DUNAI

University of Edinburgh

CAMBRIDGE
UNIVERSITY PRESS

CAMBRIDGE
UNIVERSITY PRESS

University Printing House, Cambridge CB2 8BS, United Kingdom

One Liberty Plaza, 20th Floor, New York, NY 10006, USA

477 Williamstown Road, Port Melbourne, VIC 3207, Australia

4843/24, 2nd Floor, Ansari Road, Daryaganj, Delhi - 110002, India

79 Anson Road, #06-04/06, Singapore 079906

Cambridge University Press is part of the University of Cambridge.

It furthers the University's mission by disseminating knowledge in the pursuit of education, learning and research at the highest international levels of excellence.

www.cambridge.org
Information on this title: www.cambridge.org/9781108445726

First published 2010
First paperback edition 2017

A catalogue record for this publication is available from the British Library

Library of Congress Cataloging in Publication data
Dunai, T. J. (Tibor), 1965–
Cosmogenic nuclides : principles, concepts and applications in the earth surface sciences / Tibor Dunai.
p. cm.
ISBN 978-0-521-87380-2 (hardback)
1. Cosmogenic nuclides. 2. Isotope geology. 3. Earth–Surface.
4. Cosmic rays. 5. Cosmochemistry. I. Title.
QE501.4.N9D864 2010
551.9–dc22

2009045373

ISBN 978-0-521-87380-2 Hardback
ISBN 978-1-108-44572-6 Paperback

Contents

Preface

Cosmogenic nuclides have become a widely used tool to address scientific questions in Earth surface sciences. Major advances in analytical sensitivity, accuracy and precision in the late 1980s made application to problems in Earth sciences feasible. In particular, widespread use of *in-situ*-produced cosmogenic nuclides has revolutionized Earth surface sciences in the last 15 years. The capabilities to quantify the geomorphic stability of surfaces exposed to cosmic rays and to determine long-term erosion rates were quickly adopted to address, and resolve for the first time, a wide range of first-order problems in the fields of geomorphology, glaciology, palaeoclimatology, palaeoseismology, soil science, volcanology and geohazard research. In the pioneering days of cosmogenic nuclide methodology, it was commonly the same researchers who developed, as well as applied, the methodology; with increasing specialization and division of work, and with new researchers entering the fields as users, this is becoming relatively rare. While it is not feasible, and probably not necessary, for every user of the cosmogenic methodology to know *every aspect* of the methodological basis, a firm knowledge of the fundamentals is crucial for applying the method safely in the natural environment. This is because scientific questions and field situations may often be similar, but they are rarely identical, usually requiring a knowledgeable adaptation of generic 'recipes' to design a particular scientific approach and sampling strategy. Also, readers of scientific findings based on evidence derived from cosmogenic nuclides (users *sensu lato*) should be able to assess the robustness of data in the light of the method's strengths and limitations. These user requirements have set the initial framework of this book: providing an as simple as possible and as complete as necessary account of the current state of cosmogenic nuclide methodology used in Earth surface sciences. As a consequence, this book was written with the aim to enable interested users to 'think cosmogenic', i.e. to appreciate scientific approaches used by others, as well as successfully design research applications of cosmogenic nuclides for their own research. However, I also hope that seasoned practitioners in the field will find this compendium useful.

In a textbook of this kind, it is necessary to strike a balance between the amount and type of literature cited to keep it readable, while giving sufficient guidance to readers to find their way through the plethora of pertinent literature. The literature cited must be wide ranging, but cannot be complete. I have attempted to find an equilibrium between giving appropriate acknowledgment to the pioneering studies, while, at the same time, pointing the reader in the direction of the most recent developments in the field. Also, where available for special topics, I provide references to recent review papers or papers that summarize a certain topic in an exemplary way, instead of listing the papers they rely on. I trust that I will receive feedback if I haven't got this balance right.

The past 10 years have seen a remarkable activity on aspects of cosmogenic nuclide methodology, particularity of how best to determine cosmogenic nuclide production rates for any location on Earth. At the time of writing, the research results of two international research consortia CRONUS-EU (EU-funded, 2004–2008) and CRONUS-Earth (NSF-funded, 2005–2010) indicate that one of their original goals, the ability to reproducibly obtain cosmogenic nuclide production rates better than 5%, is within reach. A wide range of additional methodological improvements (refinement of half-lives; production pathways; sampling strategies etc.) were achieved by CRONUS and other researchers world-wide over this period. As far as fitting the scope of this book, these new developments were incorporated.

This book profited inestimably from continuous discussions with fellow researchers of *CRONUS-EU* and *CRONUS-Earth*, and from exchanges with colleagues on joint cosmogenic sampling expeditions over the past 12 years. Some of these discussions were heated, and I am particularly grateful for those. The questions asked by participants of the *CRONUS-EU* workshops and summer school made me realise that there may be a gap in the literature that would be worth filling; this realization was particularly keen at moments when I struggled to give simple answers to relatively simple questions. I am very grateful for the critical eyes of my colleagues Nat Lifton, Steve Binnie, Fin Stuart and Richard Philips who have read and commented on drafts of the manuscript/chapters. They provided valuable suggestions that helped to improve the final version significantly. Any lingering omissions or factual mistakes remain my responsibility. Rachel Walcott transformed essential climate data used in Figure 3.2 into a format I could handle, which is gratefully acknowledged. Further, I am grateful for the Leverhulme research fellowship, which enabled me to create time to write this book, and for the patience of the editor Susan Francis while waiting for its delivery.

Overall, my greatest debt is to my wife Karin; without her moral support and patience, proof-reading and help with drawing and editing, the book would probably not have been finished before the publisher (and I) had lost faith in the project. This book is, therefore, for her.

Tibor J. Dunai
Edinburgh
November 2009

1
Cosmic rays

Victor Hess's experiments in manned balloons in 1912 demonstrated the existence of penetrating radiation entering the Earth's atmosphere from space. Hess's original observation was that gold-foil electrometers carried to an altitude of 5300 m indicated a rapid increase of ionization with elevation. Others confirmed this observation, and in 1925 Robert Andrews Millikan coined the term 'cosmic rays'. Hess received the Nobel Prize in Physics in 1936 for his discovery.

The term 'cosmic ray' is in fact a misnomer; it was based on the belief that the radiation was electromagnetic in nature. During the 1930s it was found that the primary cosmic rays are electrically charged – thus particulate – because they are affected by the Earth's magnetic field. Throughout the 1930s, and until the 1950s, before man-made particle accelerators reached very high energies, cosmic rays served as a unique source of particles for studies in high-energy physics, and led to the discovery of various subatomic particles, including the positron and the muon (Powell *et al.* 1959). Studies conducted during this pioneering phase laid the foundations for the theoretical and phenomenological understanding of cosmic rays as relevant for the use in Earth surface sciences today. Subsequent studies have refined this understanding; a process that is still ongoing.

1.1 Origin and nature of cosmic rays

Cosmic rays are high-energy, charged particles that impinge on the Earth from all directions. The majority of cosmic-ray particles are atomic nuclei, but they also include electrons, positrons and other subatomic

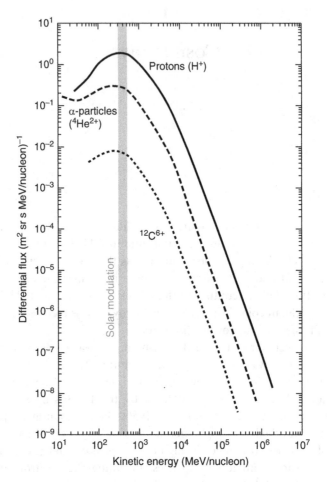

Fig. 1.1. The major components of the primary cosmic ray flux are protons and alpha-particles, the next most common particles are carbon nuclei, though only at less than <1% of the proton flux. Nuclei heavier than carbon are even rarer (not shown); data from (Simpson 1983). For a discussion of solar modulation see Section 1.2.

particles. Typical energy levels range from a few MeV up to $\sim 10^{20}$ eV, with a maximum at a few hundred MeV per nucleon (Fig. 1.1).

The term 'cosmic rays' usually refers to galactic cosmic rays, which originate in sources outside the solar system. However, this term is sometimes also used to include nuclei and electrons accelerated in association with energetic events on the Sun (solar cosmic rays). Solar cosmic rays have much lower energies (<1 GeV; typically 1–100 MeV)

The energy of primary cosmic rays

The energy of cosmic rays is usually provided in units of MeV, for mega-electron volts, or GeV, for giga-electron volts. One electron volt is the energy gained when a particle with a charge equivalent to the charge of an electron is accelerated through a potential difference of 1 volt. Most galactic cosmic rays have energies between 100 MeV (corresponding to a velocity of 43% of the speed of light for protons) and 10 GeV (corresponding to 99.6% of the speed of light). The number of cosmic rays with energies in excess of 1 GeV decreases by about a factor of 50 for every tenfold increase in energy.

than galactic cosmic rays, do not contribute significantly to the cosmogenic nuclide production at the Earth's surface (Masarik and Reedy 1995) and are therefore not considered further in the context of this book. The current understanding is that most galactic cosmic rays derive their energy from supernova explosions, which occur approximately once every 50 years in our Galaxy (Diehl *et al.* 2006). Cosmic rays are accelerated as the shock waves from these explosions travel through the surrounding interstellar gas. Sources of the primary cosmic radiation up to energies of at least 10^{15} eV are located within our galaxy (Eidelman *et al.* 2004). The mean cosmic-ray energy spectrum and integrated cosmic-ray flux is considered to be constant over the last 10 Ma (Leya *et al.* 2000).

At the top of the Earth's atmosphere the cosmic rays are largely composed of protons (87%) and α-particles (12%). A small contribution of heavier nuclei ($\sim$1%) is also present (Fig 1.1). Upon entering the Earth's atmosphere these primary cosmic rays produce secondary cosmic rays in interactions with the atoms in air.

1.1.1 The nucleonic component

The high energies of the primary cosmic rays are well in excess of the binding energies of atomic nuclei (typically 7–9 MeV per nucleon). Consequently, the predominant nuclear reaction in the atmosphere is that of spallation. In these reactions, nucleons are sputtered off the target nucleus. Spallation-produced nucleons largely maintain the direction of the impacting particle (Dorman *et al.* 1999) and go on inducing

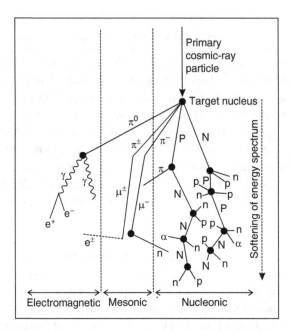

Fig. 1.2. The secondary cosmic ray cascade in the atmosphere. Abbreviations used: n, neutron, p, proton (capital letters for particles carrying the nuclear cascade), α, alpha particle, e±, electron or positron, γ, gamma-ray photon, π, pion, μ, muon. After Simpson and Fagot (1953), discussion in the text.

spallation in other target nuclides, producing a nuclear cascade in the Earth's atmosphere (Fig. 1.2). Because neutrons do not suffer ionization losses as do protons (Lal and Peters 1967), the composition of the cosmic-ray flux changes from proton-dominated to neutron-dominated in the course of the nuclear cascade. At sea level, neutrons constitute 98% of the nucleonic cosmic-ray flux (Masarik and Beer 1999). Another change in the character of the cosmic rays on their way through the atmosphere is that the energy of the secondary neutrons is significantly lower than that of primaries. At ground level, energies >10 GeV are, for the purpose of Earth-science applications, absent. The neutron energy spectrum has peaks around 100 MeV, 1–10 MeV and <1 eV (Fig. 1.3). The relative shape of the >1 MeV peaks is invariant between sea-level and high-mountain altitude (Nakamura et al. 1987, Gordon et al. 2004, Kowatari et al. 2005). The spectrum only changes at much higher elevations, on approaching 12 km (Lal and Peters 1967, Goldhagen et al. 2002).

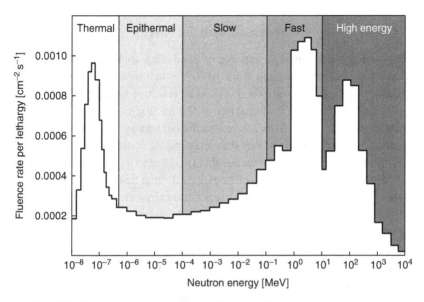

Fig. 1.3. The energy spectrum of secondary cosmic ray neutrons at sea level (data from Goldhagen *et al.* 2002). Lethargy is the natural logarithm of energy. For discussion see text.

1.1.2 The mesonic component

Collisions of high-energy primary cosmic rays with atomic nuclei high in the atmosphere produce mesons. These are mostly pions, which decay within a short distance (a few metres) predominantly to muons (μ^-, μ^+) (Eidelman *et al.* 2004) (Fig. 1.2). Muons belong to the particle family of leptons and can be considered as the heavier brother of the electron (206.7 times heavier). They have a half-life of 2.2 µs. At the speed of light this would give them a range, before they decay, of only 660 m. However, at relativistic speeds, the lifetime of the muon, as we perceive it, is much longer. Due to this time-dilation effect of special relativity, muons can reach the Earth's surface. Muons (μ^-) decay into an electron, an electron-antineutrino, and a muon-neutrino. Antimuons (μ^+) decay to a positron, an electron-neutrino, and a muon-antineutrino.

Muons are produced high in the atmosphere (typically 15 km) and lose about 2 GeV to ionization before reaching the ground. Their energy and angular distribution reflect a convolution of production-energy-spectrum, energy loss in the atmosphere and decay. For example, 2.4 GeV muons have a decay length of 15 km, which is reduced to 8.7 km by energy loss (Eidelman *et al.* 2004). The mean energy of muons at sea level is ~4 GeV

Flavours of neutron energy

High-energy neutrons are capable of producing spallation reactions.
They have energies ranging from 10 GeV down to about 10 MeV.
High-energy neutrons are themselves the result of spallation
reactions, and are the main carrier of the nuclear cascade.
Fast neutrons, 0.1–10 MeV, have insufficient energy to induce
spallation reactions. However, they may induce some evaporation
reactions, e.g. energetically favourable (n,α)* reactions. Fast neutrons
are themselves products of nuclear evaporation: a de-excitation process
named by analogy to molecules evaporating from the surface of a
heated liquid. Evaporation reactions can be induced by any particle
interaction supplying sufficient separation energy (see also Section 1.6).
Slow Neutrons, 100 eV and 100 keV, are produced continuously,
as fast neutrons lose energy through elastic and inelastic collisions
with nuclei; this process is called moderation.
Epithermal neutrons are produced by further moderation (0.5–100 eV).
Finally, **thermal neutrons** have achieved energy equilibrium with their
surroundings; their mean energy at environmental temperatures is
∼0.025 eV.

There is no generally accepted convention on the classification of
neutrons. The energy ranges defined here for the purpose of this book
are customary in cosmic-ray physics; other fields, such as nuclear
engineering, use different classifications.

Note: *A note on the notation for nuclear reactions: the first entry in the bracket denotes
the impacting particle, the second the outgoing particle. These are most commonly:
α = alpha particles (^{4}He-nuclei), n = neutrons, p = protons, γ = gamma-ray photon.

(Eidelman *et al.* 2004). Muons (and other particles) of this energy level
are generated within a cone-shaped shower, with all particles staying
within about 1 degree of the primary particle's path (Dorman *et al.*
1999).

Because muons interact only weakly with matter (mostly via ionization) they have a much longer range than nucleons, and therefore are
the most abundant cosmic-ray particles at sea level (Lal 1988). However,
as we will see later in Sections 1.4 and 1.6, the nucleonic component
dominates cosmogenic nuclide production at the Earth's surface. Muons,
in turn, dominate production at depth in the subsurface.

1.1.3 Electromagnetic component

At the Earth's surface, the electromagnetic component consists of electrons, positrons and photons, primarily from electromagnetic cascades initiated by decay of neutral and charged mesons (Fig. 1.2). Muon decay is the dominant source of low-energy electrons at sea level (Eidelman *et al.* 2004). The electromagnetic component is not relevant for the Earth science applications covered in this book and is therefore not considered further.

1.2 Interaction with magnetic fields

Primary cosmic rays, i.e. protons and α-particles, are fast-moving, positively charged particles. As is valid for any moving charged particle, magnetic and electrical fields affect primary cosmic rays. Electric fields accelerate or decelerate them in the direction of the field. In magnetic fields, the Lorentz force, F_L, accelerates charged particles radially, causing them to curve perpendicularly to both their initial vectors of movement v and the prevailing magnetic field B. The larger the angle between B and v is, the stronger the deflecting force. Furthermore, the slower the particle, i.e. the lower its kinetic energy, the more it will be deflected. As a result, low-energy particles follow intricate trajectories before reaching the Earth's surface, whereas the trajectories of high-energy particles are considerably less complex (Smart *et al.* 2000).

Primary cosmic-ray particles with energies <10 GeV are modulated by the solar wind and by the Sun's 11-year solar activity cycle (Lal and Peters 1967, Eidelman *et al.* 2004). As a consequence of this modulation, galactic cosmic-ray particles with rigidities (see text box) smaller than 0.6 GV on average (Michel *et al.* 1996) cannot approach the Earth (at present the solar modulation potential parameter ϕ ranges from 0.3–1.2 GV, depending on solar activity; Michel *et al.* 1996, Masarik and Beer 1999, Usoskin *et al.* 2005, Wiedenbeck *et al.* 2005; see also Fig. 1.1).

Near-vertically incident particles dominate the primary cosmic-ray flux near the Earth's surface (Dorman *et al.* 1999; see also Section 1.3). Consequently, primary particles approaching the Earth's geomagnetic equator travel perpendicular to the geomagnetic field, whereas near the poles they travel essentially parallel to the magnetic field lines. Virtually all rigidities are permitted at the poles, while near the equator, rigidities well in excess of 10 GV are required to approach the Earth. The solar modulation limits the lowest energies at the poles to >0.6 GV, having

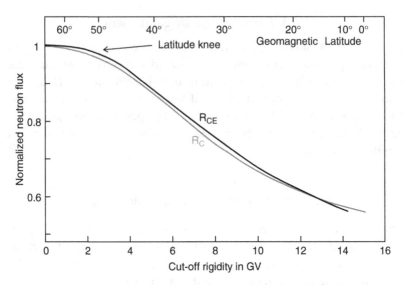

Fig. 1.4. The neutron flux at sea level varies as a function of the cut-off rigidity. The two lines show the results for two different approaches to calculate the cut-off rigidity. The curve labeled with R_{CE} was derived by trajectory tracing and the data of Rose *et al.* (1956), as derived by Desilets *et al.* (2001). The curve labeled R_C was derived using Equation 1.2 and data from Rose *et al.* (1956) from areas with a dominant dipole field (i.e. >70% dipole), as derived by Dunai (2000, 2001). To help to relate cut-off rigidity to physical locations on the globe the corresponding geomagnetic latitude is indicated as well. These were calculated for present-day field strength and should be used as an approximate guide only. See text for discussion.

a consequence that the cosmic-ray flux does not increase monotonously approaching the poles, but levels off at rigidities close to the solar modulation potential (Fig. 1.4). Furthermore primary particles with energies close to the solar modulation potential are not energetic enough to generate a secondary particle cascade that can reach the surface. The resulting break in trend at high latitudes is referred to as the 'latitude knee'. The decrease of the cosmic-ray flux with decreasing latitude below the latitude knee is sometimes referred to as the 'latitude effect'.

Approximating the Earth as a dipole field, the cut-off rigidity R_C for vertically incident particles is

$$R_C = \frac{M\mu_0 c}{16\pi RE^2} \cos^4 \lambda \ [\text{V}] \tag{1.1}$$

(Elsasser *et al.* 1956), where M is the dipole moment, μ_0 the permeability of free space, c the speed of light, RE the radius of the Earth and λ the geomagnetic

Rigidity and cut-off rigidity

Rigidity (R) is momentum per unit charge. All particles having the same magnetic rigidity, charge sign and initial conditions will have identical trajectories in the Earth's magnetic field, independent of particle mass or charge. For instance, a proton with a kinetic energy of 10 GeV and an α-particle of 5 GeV both have a rigidity of 10 GV. Cut-off rigidity is the minimum rigidity required to penetrate the Earth's magnetic field, usually presented in units of GV. Thus: $R = pc/e$ [GV], where p is the momentum of the particle [GeV/c], c is velocity of light and e the particle charge.

latitude. To enable the consideration of non-dipole components of the geomagnetic field, Eqn (1.1) can be rewritten as

$$R_C = \frac{RE}{4} \frac{Hc}{(1 + 0.25 \tan^2 I)^{2/3}} \, [V] \qquad (1.2)$$

(Rothwell 1958), where H is the horizontal field intensity and I the inclination. This analytical equation provides a phenomenological description of the cosmic-ray flux. Using Eqn (1.2) in regions with a predominant dipole field (>70% of local field), locations with the same R_C will have the same primary cosmic-ray flux ($\pm 2\%$; e.g. Dunai 2001a). Equation 1.2 being an expanded dipolar equation, it may fail to accurately predict cosmic-ray flux in regions with extreme non-dipole magnetic field anomalies, such as in the South Atlantic (Dunai 2001a, Lifton *et al.* 2005), which presently cover less than $\sim 15\%$ of the globe (www.ngdc. noaa.gov/geomag/).

Equations 1.1 and 1.2 provide a simplified picture of the energy spectrum of the primary cosmic-ray flux. This is a consequence of the solid Earth being opaque to cosmic rays, and some of the intricate trajectories at energy levels close to the cut-off rigidity intersect with the Earth and are therefore 'forbidden' (Smart *et al.* 2000). The series of allowed and forbidden rigidities for particle access near the cut-off rigidity is called the cosmic-ray penumbra (Smart *et al.* 2000). If exact high-order descriptions of the geomagnetic field are available, then numerical trajectory tracing of numerous (up to several million) modelled particles can provide an accurate image of the structure of the lowest energy levels permitted to pass the field (Smart *et al.* 2000). The mean cut-off rigidity between the first and the last permitted trajectory in the cosmic-ray penumbra defines the effective cut-off rigidity R_{CE} (Smart *et al.* 2000). The trajectory tracing-derived R_{CE} provides an accurate phenomenological description of the primary cosmic-ray flux. When using accurate descriptions of the geomagnetic field for trajectory tracing, locations with the same R_{CE} will have the same primary cosmic-ray flux ($\pm 0.3\%$; e.g. Villoresi *et al.* 1997). It is important to note that the numeric values of R_C and R_{CE} for the same location are usually different (Fig. 1.4), and they must be used consistently and not be mixed for evaluations of the cosmic-ray flux (see also Section 3.3). R_C and R_{CE} are both calculated parameters that predict the primary cosmic-ray flux, which is further modulated by the atmosphere (Section 1.3).

The flux of muons with energies above 3 GeV, i.e. those responsible for nuclear reactions pertinent for Earth surface sciences (Section 1.2), changes by less than 10% with geomagnetic latitude and/or solar activity (Stone *et al.* 1998). This lack of sensitivity to magnetic fields arises from the fact that the primary particles that produce >3 GeV muons in the upper atmosphere, which have high enough energy to reach ground level, have rigidities well in excess of 5 GV, as muons lose about 2 GeV to ionization before reaching the ground (Section 1.1). A full energy transfer from primary particle to pion (which, in turn, decays to a muon) is unlikely, as pions usually share the energy of the primary particle with other secondary particles of the nuclear cascade (Fig. 1.2). At >20 GeV, muons (or more accurately their primaries) become completely insensitive to the Earth's magnetic field (Stone *et al.* 1998).

1.3 Interactions with the Earth's atmosphere

As described earlier, the atmosphere is the location of the nuclear cascade producing secondary neutrons. After attaining a maximum of secondary neutrons at the top of the atmosphere, their abundance N decreases approximately exponentially with increasing atmospheric depth:

$$N(d) = N_0 e^{-d/\Lambda} \tag{1.3}$$

with N_0 being the number of nucleons at the top of the atmosphere, d the atmospheric depth (in units of $g\,cm^{-2}$; N.B. standard sea-level pressure

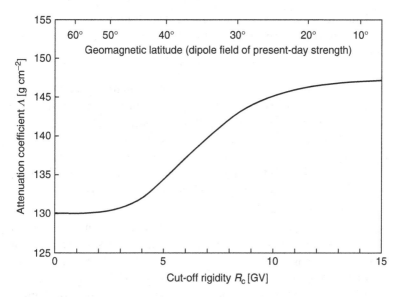

Fig. 1.5. The attenuation coefficient as a function of cut-off rigidity R_C (after Dunai 2001). The geomagnetic latitude, calculated for present-day field-strength, is provided to help to relate cut-off to physical locations on the globe.

1013 mbar $= 1030$ g cm^{-2}) and Λ the *attenuation length* (likewise in units of g cm^{-2}). As a rough guide, the nucleonic cosmic-ray flux doubles with every 1500 m increase in altitude.

The attenuation length Λ is not constant, but a function of the geomagnetic latitude λ or cut-off rigidity R_C (λ and R_C are related by Eqn (1.1); Fig. 1.5), and a function of altitude (Lal 1991, Dunai 2001b, Desilets and Zreda 2003, Desilets *et al.* 2006b). The latitude and altitude dependence of Λ is the result of the changes in the energy spectra of primary (latitude) and secondary particles (atmospheric depth; see also Sections 1.1 and 1.2).

The attenuation of muons in the atmosphere can be described by the same equation (1.3), but has a larger attenuation length Λ_μ that is a function of the muon energy: $\Lambda_\mu = (263 \pm 14) + (150 \pm 15) \cdot p$; with muon momentum p [GeV/c]; (Boezio *et al.* 2000). For atmospheric muon momenta, $p \sim 7$ GeV/c, Λ_μ is ~ 1300 g cm^{-2} (Heisinger *et al.* 2002a).

The atmospheric depth through which incoming cosmic rays must pass is thinnest for rays arriving from the zenith and increases for rays from close to horizontal. Consequently the cosmic-ray flux is greatest from the

vertical and decreases toward the horizon. The resulting angular intensity
distribution of cosmic rays is given by

$$I(\theta) = I_0 \sin^m\theta \qquad (1.4)$$

where I_0 is the vertical intensity and θ is the inclination angle measured
from the horizontal. The most commonly used value for the exponent m is
2.3 ± 0.5. Nishiizumi *et al.* (1989), and subsequently others, cite Lal (1958)
as source for the value of 2.3. However Lal (1958) actually cites Conversi
and Rothwell (1954) for the source, which gives values of 2.1 ± 0.3 and
2.6 ± 0.2 for 60 and 750 MeV nucleons, respectively. Other values used are
2.5 ± 0.5–3.0 ± 0.5 (Barford and Davis 1952); 3.5 ± 1.2 (Heidbreder *et al.*
1971) and 2.65 (Masarik *et al.* 2000).

   Neglecting the significant uncertainties for a moment, the differences in
values for m are as ought to be expected. Observations at high altitudes
(Barford and Davis 1952, Conversi and Rothwell 1954) record a less
collimated cosmic-ray flux than observations at low altitudes (Heidbreder
et al. 1971). This is because the atmospheric-depth contrast between
differently inclined pathways increases with atmospheric depth. Since
the attenuation of cosmic rays is energy dependent, m is also a function
of geomagnetic latitude (Dorman *et al.* 2000), which may account for
some of the differences. Currently the uncertainties in these observations
(see discussion in Heidbreder *et al.* 1971) are too large to draw definitive
conclusions on the nature of the discrepancies. Fortunately, the discrepancies do not affect most Earth science applications of cosmogenic
nuclides (Chapter 4).

   The angular distribution of muons follows the same relationship as that
of nucleons (Eqn. (1.4)). A value of $m = 2$ is characteristic of muons with
~ 3 GeV energy (Eidelman *et al.* 2004). At lower energy, the angular
distribution becomes increasingly steep, while at higher energy it flattens
(Eidelman *et al.* 2004); a behaviour equivalent to that of nucleons
(Dorman *et al.* 2000).

1.4 Interactions with the Earth's surface

The attenuation of cosmic rays in the solid Earth is, in principle, the same
as in the atmosphere. The main differences are the much higher density of
rocks and changes in mean atomic mass and charge per nucleus. Further,
there are boundary effects at the Earth–atmosphere interface that affect
thermal neutrons.

1.4.1 Fast and high-energy neutrons

The fast- and high-energy-neutron flux decreases approximately exponentially (Dunne *et al.* 1999) with increasing depth below the surface:

$$N(z) = N_0 e^{-z\rho/\Lambda} \tag{1.5}$$

with N_0 being the number of nucleons at the surface, z the depth below surface (cm), ρ the density of the rock (g cm^{-3}) and Λ the attenuation length (g cm^{-2}).

'True' and 'apparent' attenuation lengths

The attenuation of a vertically incident particle flux onto a horizontal surface is described by the 'true' attenuation length (Gosse and Phillips 2001). However, non-vertically incident particles also contribute to the subsurface cosmic-ray flux. The attenuation length describing the decrease of the overall cosmic-ray flux in the subsurface is called the 'apparent' or 'effective' attenuation length (Brown et al. 1992, Gosse and Phillips 2001). The numeric value of the 'true' attenuation length is about 1.3 times larger than that of the 'apparent' (Gosse and Phillips 2001). The apparent attenuation length is usually the pertinent length to describe cosmic-ray flux as relevant to Earth surface science applications and is consistently used throughout this book, without prefix.

The attenuation length of high-energy neutrons ranges between 150 and 180 g cm^{-2} (Kurz 1986b, Brown *et al.* 1992, Sarda *et al.* 1993, Nishiizumi *et al.* 1994, Gosse and Phillips 2001, Farley *et al.* 2006, Farber *et al.* 2008). Values are lower at high latitudes (Brown *et al.* 1992) than at mid- and low-latitude sites (all others). This is a similar effect as observed for the atmosphere (Fig. 1.6) and is a consequence of the increasing cosmic-ray rigidity with decreasing latitude (Fig. 1.4). Generally, higher-energy particles penetrate deeper than low-energy particles. Some variation of values for Λ may arise from contrasts in composition and differences in the mean energy of the nucleons observed.

For a granitic rock with a density of 2.65 g cm^{-3} the *attenuation path length* for high-energy neutrons is 57–68 cm (using the above range for Λ and Eqn. (1.5); the attenuation path length is the distance over which the cosmic-ray flux decreases by a factor of $1/e$, or 63%). This value decreases with increasing density and vice versa. For basalts, the attenuation path length

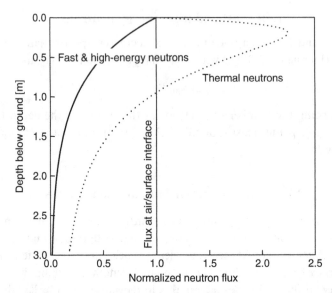

Fig. 1.6. Neutron flux below the air–surface interface. The fast- and high-energy-neutron flux decreases exponentially with depth. Thermal neutrons that are created near the interface can leak back into the atmosphere, causing the humped profile. The location of the hump depends on the composition of the rock and its water content (Phillips *et al.* 2001). The flux profiles were calculated using CHLOE (Phillips and Plummer 1996) using $\rho = 2.5\,\mathrm{g\,cm^{-3}}$ and average granitic composition (Andrews and Kay 1982), with 3% water. The fluxes were normalized to the corresponding flux at the air–surface interface.

is $\sim$50 cm. At the depth of five attenuation path lengths, i.e. 2.5–3 m in most rocks, <1% of the high-energy-neutron flux arriving at the surface remains (Fig. 1.6).

1.4.2 Thermal neutrons

The majority of cosmic-ray neutrons are moderated to lower energies by inelastic scattering by nuclides in the atmosphere or in rocks. Due to the high absorption cross-section of nitrogen for thermal neutrons, the corresponding cosmogenic thermal neutron (CTN) flux in rocks is approximately one order of magnitude higher than in the atmosphere (Phillips *et al.* 2001). Therefore, the CTN flux in rocks is directly derived from high-energy cosmic-ray neutrons subsequently moderated in the rocks (Liu *et al.* 1994a, Phillips *et al.* 2001). Consequently the CTN flux in rocks co-varies with the high-energy-neutron flux as a function of altitude and latitude (see Sections 1.2 and 1.3).

Contrary to the highly directional downward flux of high-energy neutrons, the movement of moderated low-energy neutrons is random (Phillips *et al.* 2001). This together with the long mean attenuation path length of neutrons in rocks ($\sim$50 cm in granite, up to $\sim$75 cm in ultramafic rocks; $\Lambda_{thermal} \approx 23$ g cm^{-2}; Andrews and Kay 1982, Lal 1987, Liu *et al.* 1994a) causes diffusive loss (also termed 'leakage') of neutrons to the atmosphere (Phillips *et al.* 2001), where they are quickly absorbed by nitrogen. This leakage at the rock–atmosphere interface results in a maximum CTN flux positioned at a depth of approximately 50 g cm^{-2} (Phillips *et al.* 2001); this corresponds to $\sim$20 cm below the surface (Fig. 1.6). The CTN flux at the bulge maximum is approximately twice the flux at the rock–atmosphere interface (Phillips *et al.* 2001). Hydrogen, in soil moisture or lattice water, effectively moderates neutron energies and has a significant influence on the magnitude and the depth of the CTN flux maximum. Water in rocks generally increases the CTN flux by up to 20–25% (Phillips *et al.* 2001). The largest increase occurs at approximately 5% water in granitic rocks (Phillips *et al.* 2001). With increasing water content, the CTN flux maximum moves upward to the rock–atmosphere interface (Phillips *et al.* 2001). Consequently thin snow cover and/or water-saturated soil/moss reflect thermal neutrons that would otherwise leak to the atmosphere and increase the thermal-neutron flux at the surface. The effect of snow and moist soil above rock cannot be modelled with currently available programs (Phillips *et al.* 2001).

1.4.3 Muons

Due to their weak interaction with matter, muons penetrate deeper into the subsurface than neutrons. Muons are dominantly slowed by ionization at a rate of 2.0 ± 0.3 MeV g cm^{-2} (for muons 100–50 GeV; Groom *et al.* 2001), i.e. they lose about 1 GeV for every 2 m in rocks, and eventually come to rest. At 4 GeV, the mean energy of muons at the Earth's surface (Eidelman *et al.* 2004), ionization accounts for 99.75% of the deceleration (Groom *et al.* 2001). At very high energies, other mechanisms, such as Bremsstrahlung, become important, and contribute $\sim$10% to the deceleration of 100 GeV muons (Groom *et al.* 2001). Average muon ranges for standard rock (density 2.65 g cm^{-3}, average

atomic number (Z) and nucleon number (A) of 11 and 22, respectively;
Groom *et al.* 2001) are:

     1 GeV   2 m
     4 GeV   7.9 m
   10 GeV   19 m
 100 GeV   154 m

The mean attenuation length of these stopping negative muons, $\Lambda_{\mu-}$, is 1510 g cm^{-2} (Heisinger *et al.* 2002a).

Fast muons give rise to Bremstrahlung of sufficiently high energy to produce secondary neutrons via (γ,n) reactions, so-called photoneutrons (Groom *et al.* 2001). The attenuation length for nuclear reactions caused by these photoneutrons, $\Lambda_{\mu,\text{fast}}$, is 4320 g cm^{-2} (Heisinger *et al.* 2002b).

The strong influence of the chemical composition on muon attenuation (Groom *et al.* 2001) is the reason for the difference in attenuation in air (Section 1.3) and the subsurface. To describe muon stopping and attenuation in rocks with non-standard composition (see above; Groom *et al.* 2001) may require different attenuation lengths than those given above.

1.5 Production of cosmogenic nuclides

Cosmogenic nuclides are the products of interactions of primary and secondary cosmic-ray particles with atomic nuclei. At the Earth's surface, more than 98% of the cosmogenic nuclide-production arises from secondary cosmic-ray particles (Masarik and Beer 1999), such as neutrons and muons (Section 1.1). Depending on the energy of these particles, a range of nuclear reactions produce cosmogenic nuclides.

1.5.1 Spallation

In spallation reactions, high-energy neutrons (at ground level rarely protons) collide with atomic nuclei ('target nuclei') and sputter off protons and neutrons, leaving behind a lighter nucleus. These reactions take the form of an intranuclear cascade (Serber 1947, Filges *et al.* 2001) where during a series of independent nucleon–nucleon collisions, the nucleons act like billiard balls. If the impact is peripheral, several secondary nucleons can escape with much of the impactor's energy (10–100 MeV) and continue the nuclear cascade. If the impact is central, the impactor's energy can be distributed over the nucleus, leaving it in an excited state.

If the excited state of the nucleus has an energy of more than 7–9 MeV (the typical binding energy of a nucleon) above the ground state, subsequent de-excitation of the nucleus can result in evaporation reactions emitting one or more nucleons with low MeV energy (Serber 1947, Masarik and Beer 1999). The mean energy of evaporation neutrons is about 1 MeV (Fig. 1.3).

Due to the large binding energy of α-particles (28.3 MeV, or 7.075 MeV per nucleon) their emission is energetically favoured over the alternative emission of the four individual nucleons. A neutron of <5 MeV can trigger an (n,α) reaction, but will be unable to trigger a single (n,n) or (n,p) reaction. Generally, reactions that involve the emission of an α-particle have large *reaction cross-sections* δ (cross-sections express the likelihood of interaction between particles, they have the unit of area, 1 barn $= 10^{-28}$ m; analogous to 'a large target is more likely to be hit'). Similar, but much less pronounced energy advantages are associated with the emission/sputtering of ^{3}He, ^{3}H (tritium) and ^{2}H (deuterium).

The mass difference between a target nucleus and the lighter product is usually a few atomic mass units. Thus, generally, target elements with nuclide masses nearest to the cosmogenic nuclide produced contribute most to its production. A notable exception is ^{3}He that is, together with its precursor ^{3}H (^{3}H$(\beta^-)^3$He; half-life 12.32 years; Lucas and Unterweger 2000), sputtered off from nuclei of all masses and does not represent a residual nucleus. Examples for relevant spallation reactions producing cosmogenic nuclides are given in Table 1.1.

The reactions producing the various cosmogenic nuclides each have their specific *energy threshold Q*, above which they can occur, and have energy-dependent reaction cross-sections (Fig. 1.7). They therefore respond to different energies in the neutron energy spectrum (Fig. 1.3).

1.5.2 Thermal neutron capture

The majority of neutrons of the nuclear cascade are slowed down to become thermal neutrons (Phillips *et al.* 2001). Some thermal-neutron capture reactions have very large reaction cross-sections and can produce appreciable amounts of cosmogenic nuclides. Relevant reactions for Earth surface science applications are ^{6}Li$(n,\alpha)^3$H$(\beta^-)^3$He; ^{14}N$(n,p)^{14}$C; ^{35}Cl$(n,\gamma)^{36}$Cl and ^{39}K$(n,\alpha)^{36}$Cl (Lal 1987, Phillips *et al.* 2001, Dunai *et al.* 2007). The energy of these reactions cannot be supplied by the captured thermal neutrons, as they have an energy of only $\sim$0.025 eV, but is sourced by the lower total nucleon binding energy of the reaction

Table 1.1.

Isotope	Primary spallation reactions	Thermal neutron capture	Negative muon capture
^{3}He ^{3}H(β^{-3}He)	On all elements heavier than itself, reactions of the type: (n,x^{3}He) or (n,x^{3}H) with x being any other particle or none	^{6}Li(n,α)^{3}H (β^{-3}He)	On all elements heavier that itself; reactions of the type: (μ^-,^{3}He) or (μ^-,^{3}H).
^{10}Be	^{16}O(n,^{3}Heα)^{10}Be or ^{16}O(n,4p3n)^{10}Be ^{28}Si(n,x)^{10}Be	^{9}Be(n,γ)^{10}Be	^{16}O(μ^-,αpn)^{10}Be ^{28}Si(μ^-,x)^{10}Be
^{14}C	^{16}O(n,2pn)^{14}C ^{28}Si(n,x)^{14}C	^{17}O(n,α)^{14}C ^{14}N(n,p)^{14}C	^{16}O(μ^-,pn)^{14}C
^{21}Ne	^{23}Na(n,^{3}H)^{21}Ne ^{24}Mg(n,α)^{21}Ne ^{27}Al(n,α^3H)^{21}Ne ^{28}Si(n,2α)^{21}Ne		^{23}Na(μ^-,2n)^{21}Ne ^{24}Mg(μ^-,^{3}H)^{21}Ne ^{27}Al(μ^-,α2n)^{21}Ne
^{22}Ne	^{23}Na(n,pn)^{22}Ne ^{25}Mg(n,α)^{22}Ne ^{27}Al(n,αpn)^{22}Ne ^{29}Si(n,2α)^{22}Ne		^{23}Na(μ^-,n)^{22}Ne ^{24}Mg(μ^-,pn)^{22}Ne ^{27}Al(μ^-,αn)^{22}Ne
^{26}Al	^{28}Si(n,2np)^{26}Al		^{28}Si(μ^-,2n)^{26}Al
^{36}Cl	^{39}K(n,α)^{36}Cl ^{40}Ca(n,αn)^{36}Cl	^{35}Cl(n,γ)^{36}Cl ^{39}K(n,α)^{36}Cl	^{39}K(μ^-,^{3}H)^{36}Cl ^{40}Ca(μ^-,α)^{36}Cl
^{53}Mn	^{54}Fe(n,pn)^{53}Mn ^{56}Fe(n,3np)^{53}Mn		^{54}Fe(μ^-,n)^{53}Mn

products. It is notable that (n,α) reactions are common due to the particularly large energy 'bonus' of one of the products.

Thermal neutrons are not exclusive to the nuclear cascade, but are also produced within rocks by (α,n) reactions involving light elements (Andrews and Kay 1982), the source of the α-particles being radioactive elements such as uranium, thorium and to a lesser extent samarium. These thermal neutrons can give rise to the same reactions as cosmogenic thermal neutrons. Near the Earth's surface, however, the cosmogenic thermal-neutron flux far exceeds the (α,n)-produced neutron flux (Dunai *et al.* 2007). The latter can be important for stable products (such as ^{3}He) that can accumulate over the geological age of the rock (Andrews and Kay 1982, Lal 1987, Farley *et al.* 2006, Dunai *et al.* 2007).

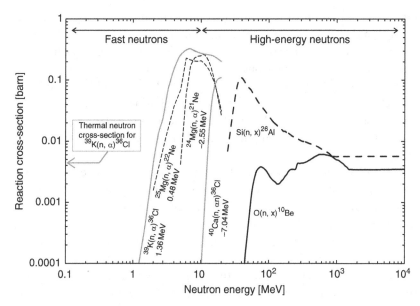

Fig. 1.7. Reaction cross-sections as a function of incident neutron energy. For ^{36}Cl- and ^{21,22}Ne-producing reactions Q-values (see text) are provided as well. Production of ^{36}Cl via the (n,α) reaction on ^{39}K from thermal neutrons is significant (cross-section is indicated). The corresponding thermal neutron reaction on ^{25}Mg producing ^{21}Ne with a lower, but still positive Q-value, has a cross-section of $<10^{-10}$ barn, i.e. is not significant (not shown). Reactions with negative Q-values cannot be induced by thermal neutrons. Data sources: ^{10}Be and ^{26}Al (Leya *et al.* 2000); ^{21,22}Ne-producing reactions ENDF/B-VII.0 database. EXFOR and ENDF databases were accessed via Janis (http://www.nea.fr/janis/). The energy regions indicated by the arrows correspond to those in the secondary cosmic-ray energy spectrum shown in Fig. 1.3.

1.5.3 Negative muon capture

Decelerated (stopped) negative muons of thermal energy can be captured by the electron shell of atoms, and quickly cascade to the lowest electron shell. There they either decay, or are captured by the nucleus. The latter is facilitated by the fact that the heavier muons have orbits much nearer to the nucleus than electrons (Eidelman *et al.* 2004). Captured muons neutralize one proton, and deliver 106 MeV (the rest mass of a muon) to the nucleus. Much of this energy is carried away by neutrino emission, such that usually 15 MeV are available for nuclear evaporation (Charalambus 1971, Stone *et al.* 1998). Due to the relatively low excitation energy, emission of α-particles, ^{2}H, ^{3}H and ^{3}He nuclei are favoured over the random evaporation of several nucleons. Relevant reactions producing cosmogenic nuclides are presented in Table 1.1.

Cross-sections for negative muon capture reactions are much lower than those involving neutrons. Consequently negative muon capture reactions account for 2%, 2.1% and 10% of ^{10}Be, ^{26}Al and ^{14}C produced, respectively (at sea level and high latitude; Heisinger *et al.* 2002a), the remainder being produced by spallation reactions involving high-energy neutrons. A similar range applies to reactions producing ^{36}Cl (Heisinger *et al.* 2002a).

This relatively small contribution of muon capture decreases with increasing altitude; because cosmic-ray nucleons have a shorter attenuation length than muons, i.e. the nucleon flux increases faster with altitude than the muon flux (Section 1.3).

For the equivalent reason the importance of muogenic reactions increases in the subsurface (Section 1.4). The nucleon flux decreases rapidly with depth below the Earth's surface, whereas muons can penetrate deeply into the subsurface (Section 1.4). For instance, at 3 m depth, the muogenic production of ^{10}Be has already overtaken spallogenic production (Fig. 1.8). At greater depth, muogenic reactions remain as the only relevant reactions producing cosmogenic nuclides.

1.5.4 Fast muons

Fast muons give rise to Bremstrahlung of sufficiently high energy to produce secondary neutrons via (γ,n) reactions, so-called photoneutrons (Groom *et al.* 2001; Section 1.4). Depending on their energy, these neutrons can cause the entire range of neutron reactions described above. Their abundance is generally very low and they only become important at great depth when fast muons remain as the sole reaction-inducing 'survivors' of the cosmogenic cascade (Fig. 1.8).

1.6 Detection of cosmic rays

Most of our knowledge of cosmic rays, particularly their flux as a function of atmospheric depth and magnetic field strength, relies on observations made with cosmic-ray detectors. In the course of the scientific exploration of cosmic rays, electrometers, Geiger counters and cloud chambers have been used to describe the properties of cosmic rays. In the late 1930s and 1940s, photographic emulsions (Blau and Wambacher

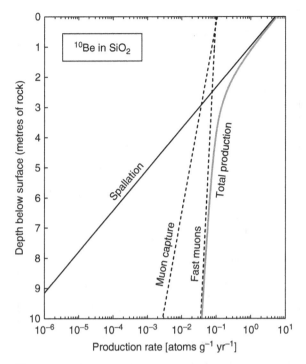

Fig. 1.8. ^{10}Be production as a function of depth below surface at sea level and high latitude. Calculated using a rock density of $2.7\,g\,cm^{-3}$, depth dependency and surface production rates from Heisinger *et al.* (2002a, 2002b).

1937, Powell *et al.* 1959) and neutron monitors (Agnew *et al.* 1947, Simpson 1951, Hatton 1971) came into use. The bulk of the phenomenological description of the cosmic-ray flux and their energies was derived with these latter modes of detection. Consequently, photographic emulsions and/or neutron monitor data form the backbone of all currently used methods to describe cosmic-ray flux as relevant to Earth surface sciences (Lal 1991, Dunai 2000, Stone 2000, Dunai 2001a, Desilets and Zreda 2003, Balco *et al.* 2008, Lifton *et al.* 2008). Their working principles and characteristics are outlined in the following.

1.6.1 Photographic emulsions

Light and ionizing radiation can cause reduction of silver ions to silver atoms in silver salts by dislocation of electrons (Powell *et al.* 1959). The individual silver atoms, or clusters of atoms, carry the latent image.

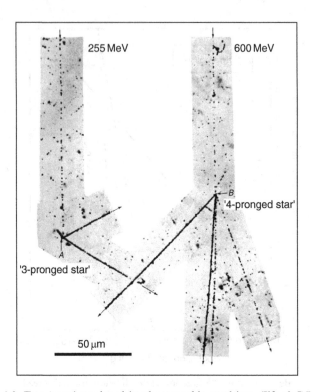

Fig. 1.9. Two 'stars' produced in photographic emulsions (Ilford G5), the 3-pronged star was produced by a proton of 255 MeV energy, the 4-pronged star by a proton of 600 MeV. Note that high-energy particles induce weaker tracks than lower-energy particles, and only charged particles produce tracks. Prior to the advent of neutron monitors the overwhelming majority of cosmic-ray research was conducted using photographic emulsions; they allow detailed analysis of individual nuclear reactions. (Based on Plate 13–3 of Powell *et al.* 1959).

The latent image can be developed by electrochemical reactions that transform the entire silver-salt crystal into visible clusters of metallic silver. This process is the basis for analogue photography.

Cosmic rays can produce nuclear reactions within the photographic emulsions (usually silver halide suspensions in gelatine). Charged, ionizing products of these nuclear reactions are recorded as tracks in the emulsion. Neutrons do not leave tracks in emulsions; however, the products of neutron reactions can leave tracks, if they are charged. Track lengths and the number of tracks ('prongs') of a 'star' provide information on the energy of the nuclear reaction recorded (Powell *et al.* 1959) (Fig. 1.9). High-energy reactions can cause stars with many prongs, and/or high-energy secondary particles. High-energy particles form long, but weak, tracks (due to weak interaction), low-energy particles form short, dark tracks (Powell *et al.* 1959).

In cosmic-ray surveys, the abundance of stars, differentiated by their number of 'prongs', has been used to derive the altitude dependence of the cosmic-ray flux as a function of the cosmic-ray energy (Teucher 1952, Dixit 1955, Lal 1958).

1.6.2 Neutron monitor

Geiger counting tubes (proportional counting tubes) filled with BF_3 or ^{3}He measure the flux of thermal neutrons by recording the ionization pulses caused by neutron capture reactions, i.e. $^{10}B(n,\alpha)^7Li$ ($+2.4$ MeV) or $^3He(n,p)^3H$ ($+764$ keV), respectively. These reactions have very large cross-sections for thermal neutrons (3840 barn and 5330 barn, respectively; Clem and Dorman, 2000) and are relatively insensitive to higher-energy neutrons (Clem and Dorman 2000).

In order to be able to detect neutrons of higher energies, neutron monitors have moderators (made of paraffin or polyethylene; the hydrogen in these hydrocarbons acts as the main moderator) around the neutron counting tubes. At its simplest, the moderator slows down fast neutrons to thermal energies *and* at the same time suppresses environmental thermal neutrons (i.e. neutrons that already have thermal energies before entering the moderator) from reaching the counting tube (Vega-Carillo *et al.* 2005). The thicker the moderator is, the higher the mean energy of recorded neutrons (Vega-Carillo *et al.* 2005).

To increase counting rate, and thus improve counting statistics, most neutron monitors used for cosmic-ray surveys and in cosmic-ray observatories also have a thick layer of lead surrounding the counting tubes. Due to the high density of the lead, incoming cosmic-ray particles (mostly neutrons, but also protons and muons) have a high likelihood of interacting with the lead and producing secondary neutrons. The secondary neutrons are reflected by the surrounding moderator and recorded by the counting tube (Hatton 1971, Clem and Dorman 2000). The standard IGY and MN64 neutron monitors have an additional moderator between the lead neutron-producer and the counting tube to further enhance the counting rate (Hatton 1971). The response functions of IGY and NM64 monitors are different, mostly because of the varying thickness of the external moderator (Hatton 1971, Clem and Dorman 2000), with the median energies of events counted by an IGY or NM64 monitor being

160 ± 40 MeV and 130 ± 30 MeV, respectively (Hatton 1971). The IGY and NM64 neutron monitors are designed for high counting rates (Simpson 1958, Hatton and Carmichael 1964), but remain relatively crude instruments with regard to resolving energy spectra (Hatton 1971, Clem and Dorman 2000).

To obtain a detailed image of the neutron energy spectrum (Fig. 1.3), sets of spherical neutron monitors, with polyethylene spheres of varying thickness as moderators, are used (Bramblett *et al.* 1960, Vega-Carillo *et al.* 2005). These sets are called 'Bonner spheres' (Bramblett *et al.* 1960). Due to their low counting rates they are rarely, and only recently, used for cosmic-ray surveys (e.g. Goldhagen *et al.* 2002, Kowatari *et al.* 2005). Systematic measurements with Bonner spheres could reveal even subtle changes of the neutron energy spectrum as a function of altitude (Goldhagen *et al.* 2002, Gordon *et al.* 2004, Kowatari *et al.* 2005) and latitude. The available data (Goldhagen *et al.* 2002, Gordon *et al.* 2004, Kowatari *et al.* 2005) already provides valuable information. For instance, the neutron energy spectrum >1 MeV, i.e. the energies relevant for fast-neutron and spallation reactions, probably does not change significantly with altitude between sea-level and high-mountain altitude (Section 1.1).

A more complete knowledge of the incident neutron energy spectrum would be valuable for improving and/or testing methods describing cosmic-ray flux as relevant to Earth surface sciences. All currently used scaling factors (Lal 1991, Dunai 2000, Stone 2000, Dunai 2001a, Desilets and Zreda 2003, Desilets *et al.* 2006, Balco *et al.* 2008, Lifton *et al.* 2008; see Chapter 3) implicitly assume that cosmogenic nuclide production rates scale linearly to the behaviour of neutron monitors (IGY or NM64) and/or photographic emulsions.

2

Cosmogenic nuclides

As early as 1934 it was hypothesized that cosmic rays could produce nuclides by interaction with Earth surface materials (Grosse 1934). In the late 1940s, radiocarbon produced by cosmic rays in the atmosphere was discovered, and developed into a widely used tool for dating organic matter (Libby 1946, Libby *et al.* 1949). The discovery and application of *in situ*-produced cosmogenic nuclides in terrestrial material took off sluggishly. ^{36}Cl was the first *in situ*-produced nuclide detected in rocks (Davis and Schaeffer 1955). At that time Davis and Schaeffer (1955) also developed the main methodological principles for exposure dating. It took more than 20 years before the next *in situ* study reported cosmogenic Xe (^{124}Xe, ^{128}Xe and ^{131}Xe) in barite from southern Africa (Srinivasan 1976), followed by measurements and production rate estimates of cosmogenic nuclides derived from airplane wreckage (Yokoyama *et al.* 1977). In the meteoritic and lunar scientific community, meanwhile, cosmogenic nuclides have been a routinely used research tool since the 1960s (Reedy *et al.* 1983, Wieler 2002). Also, atmospheric and oceanic sciences have utilized cosmogenic nuclides since that time (Lal *et al.* 1958, Lal and Peters 1962, 1967). This early adoption of the 'cosmogenic tool' was mostly due to the fact that cosmogenic nuclide production rates in extraterrestrial material and the atmosphere are several orders of magnitude higher than at the Earth's surface, and therefore detection was analytically feasible at an earlier stage.

In 1986/87 terrestrial *in situ* cosmogenic nuclides finally came of age. Eight seminal papers reported cosmogenic ^{3}He, ^{21}Ne, ^{22}Ne, ^{10}Be, ^{26}Al and ^{36}Cl in terrestrial rocks (Craig and Poreda 1986, Klein *et al.* 1986, Kurz 1986b, a, Nishiizumi *et al.* 1986, Phillips *et al.* 1986, Marti and Craig 1987, Nishiizumi *et al.* 1987). The timing of the discovery of *in situ* ^{10}Be and ^{26}Al and the revival of ^{36}Cl coincided with a major analytical improvement – the development of accelerator mass spectrometry (AMS) – which made the measurements possible (cf. Hellborg and Skog 2008).

In the following, 'terrestrial *in situ*-produced cosmogenic nuclides' (TCN; Gosse and Phillips 2001) will be termed simply 'cosmogenic nuclides', for readability, and because extraterrestrial cosmogenic nuclides are not covered in this book; confusion is therefore unlikely. Cosmogenic nuclides produced in the Earth's atmosphere, which are discussed as interference for *in situ* cosmogenic nuclide measurements, will be addressed as 'atmospheric' or 'meteoric'. They are also sometimes referred to as 'garden variety' in the literature (Nishiizumi *et al.* 1986).

2.1 'Useful' cosmogenic nuclides

From the physical principles outlined in the first chapter, it is apparent that many kinds of nuclides are produced by cosmic rays in rocks at or near the Earth's surface. Cosmogenic nuclides produced by spallation can have any mass that is smaller than the available target nuclei; those produced by thermal neutrons *can* be heavier (by one atomic mass unit) than the capturing nuclei. To be useful for Earth-science applications, however, a cosmogenic nuclide has to meet several conditions:

(i) The first condition is that it is naturally rare in geological material. Ideally it would not occur in rocks if not produced by cosmic rays. It is necessary to resolve the relatively few atoms produced by cosmic rays from the natural background concentration. The radioactivity of a nuclide, i.e. the natural depletion of a background with time, can help to achieve this condition.

(ii) The second condition is that it is either stable or long-lived radioactive. The half-life should be of the same order, or greater than, the duration of the geological process investigated.

(iii) The third condition is that its naturally occurring interferences, e.g. a meteoric cosmogenic component or any geological background, can be resolved analytically.

(iv) The fourth condition is that there is a reasonable understanding of the production mechanisms of the nuclide, e.g. knowledge of the relevant target element(s), and the relative contributions of spallation, thermal neutrons and muons to the nuclide's production.

(v) The fifth condition is that the analytical effort is feasible; not all nuclides meeting the above conditions can actually be analysed with confidence and/or reasonable effort.

(vi) The sixth condition is that the cosmogenic nuclide should be produced and retained in reasonably common minerals. The definition of 'reasonably common' is up to the judgement of potential users, and should not preclude valuable specialist applications.

Applying the above conditions, there remains a relatively short list of 'useful' nuclides for the 'cosmogenic toolbox'. These are the stable, rare noble gas isotopes: ^{3}He, ^{21}Ne, ^{22}Ne, ^{36}Ar, ^{38}Ar, various Kr and Xe-isotopes, and the radio-nuclides: ^{10}Be, ^{14}C, ^{26}Al, ^{36}Cl, ^{41}Ca and ^{53}Mn. Of these, ^{3}He, ^{21}Ne, ^{22}Ne, ^{10}Be, ^{26}Al, and ^{36}Cl are routinely used in Earth-science applications. The others are developed to a varying degree of maturity (^{14}C, ^{36}Ar, ^{38}Ar, ^{41}Ca and ^{53}Mn), or await renaissance for specialist applications (cosmogenic Xe) or pioneering development (cosmogenic Kr). The characteristics and peculiarities of these 'useful' nuclides will be addressed in Sections 2.2 and 2.3. An overview of currently used nuclides is given in Table 2.1.

Before discussing each nuclide separately, however, the fundamental differences between how stable and unstable cosmogenic nuclides accumulate at or near the Earth's surface will be introduced. Assuming that the production rate is constant with time, and that a given target is continuously exposed, the concentration of stable cosmogenic nuclides increases monotonously at a constant rate (Fig. 2.1). There is no theoretical upper limit for the exposure time that could be resolved when analysing a stable nuclide concentration, provided the production rate is known. Cosmogenic radionuclides, however, start to decay as soon as they are produced. Their specific decay constant (λ, unit [s^{-1}]), and the number of atoms available for decay, determine the rate (atoms s^{-1}) at which they decay. After an exposure time equivalent to about 2–3 times the radionuclide's half-life ($T_{1/2} = \ln(2)/\lambda$), the rate of radioactive decay and the rate of cosmogenic production become similar and the concentration of the nuclide approaches a *secular equilibrium* or *saturation*, i.e. becomes time invariant. No information on the exposure duration, other than that the minimum amount of time required to reach saturation has elapsed, can be obtained from cosmogenic radionuclides at or near saturation. Therefore, the time span over which exposure duration can be resolved with cosmogenic radionuclides is usually limited to about 2–3 times their half-lives

Table 2.1.

Isotope (half-life)	Main target minerals	Predominant target elements	Reaction pathways (SLHL)
^{3}He (stable)	Olivine, Pyroxene, other He-retentive minerals	All major elements and Li	Spallation: 100% Muons: negligible Thermal neutrons produce ^{3}He on Li, via precursor ^{3}H ($T_{1/2} = 12.3$ a)
^{10}Be (1.36 ± .07 Ma)	Quartz (rarely Pyroxene and Olivine)	O, Si (Mg)	Spallation: 96.4% Muons: 3.6%
^{14}C (5730 ± 30 a)	Quartz	O, Si	Spallation: 82% Muons: 18%
^{21}Ne, ^{22}Ne (stable)	Quartz, Pyroxene, Olivine	Mg, Al, Si	Spallation: >96.4% Muons: ≤3.6%
^{26}Al (708 ± 17 ka)	Quartz	Si	Spallation: 95.4% Muons: 4.6%
^{36}Cl (301 ± 2 ka)	Carbonates, Feldspar, Whole rock	K, Ca, Cl (Fe, Ti)	K: spallation 95.4%; muons 4.6% Ca: spallation 86.6%; muons 13.4% Fe, Ti: spallation presumed 100% Thermal neutrons produce ^{36}Cl from Cl and K.
^{36}Ar, ^{38}Ar (stable)	Feldspar, Amphibole, Pyroxene	K, Ca,	Spallation: up to 100% Muons: not determined Thermal neutrons produce ^{36}Ar from Cl and K, via precursor ^{36}Cl ($T_{1/2} = 301$ ka).
^{41}Ca (104 ± 4 ka)	Fe–Ti oxides	Fe, Ti, (Ca)	Fe, Ti: spallation 100% Thermal neutrons produce ^{41}Ca on ^{40}Ca
^{53}Mn (3.7 ± .4 Ma)	Fe-bearing minerals	Fe, Mn	Fe: spallogenic 90.2%; muons 9.8% Mn: not determined.

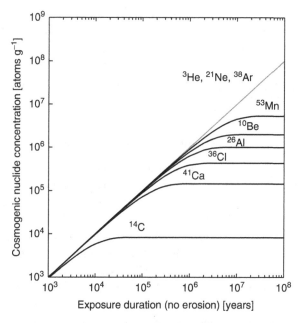

Fig. 2.1. Accumulation of cosmogenic nuclides in a non-eroding surface. Radioactive nuclides approach a secular equilibrium between production and decay after 2–3 half-lives; this situation is also referred to as a nuclide being saturated. The concentration of stable nuclides increases continuously. The curves are calculated for a production rate of 1 atom g^{-1} year^{-1}; the production rate has an influence on the level of the saturation concentration but not on the time that is necessary to achieve it.

(Fig. 2.1). The actual limit is dependent on the analytical accuracy and precision achieved.

2.2 Stable cosmogenic nuclides

The useful stable cosmogenic nuclides are the rare noble-gas isotopes. A low geological background concentration of noble gases is commonly achieved due to the fact that they are highly incompatible elements; i.e. they preferentially fractionate into the gas or the liquid phase and are not readily incorporated into minerals (Porcelli *et al.* 2002). Due to the stable nature of noble gases, any isotopes present at the time of crystallization, or acquired during the geological history of a rock, will remain in the mineral, if it is not lost by diffusion. These inherited noble-gas components can be trapped in fluid inclusions and/or be produced by nuclear reactions in the rock. A significant part of the analytical effort is often aimed at resolving these components.

Not all minerals quantitatively retain cosmogenic noble gases. Consequently, the retentivity requires evaluation for each target mineral/nuclide couple used.

2.2.1 Cosmogenic helium (^{3}He)

Helium-3 is one of the two stable isotopes of helium, ^{3}He and ^{4}He, and is a very rare isotope of a rare element. The ^{3}He/^{4}He ratio of atmospheric He is $1.39 \cdot 10^{-6}$, and there is as little as 5.24 ppm helium (by volume) in air (Mamyrin *et al.* 1970, Porcelli *et al.* 2002). Geochemical reservoirs in the Earth's mantle have 5–50 times higher ^{3}He/^{4}He ratios than air, whereas crustal rocks normally have values 10 to 100 times lower than air (the atmospheric or air ratio is customarily used as a reference value for terrestrial samples) (Ballentine and Burnard 2002, Dunai and Porcelli 2002, Graham 2002, Stuart *et al.* 2003). Pure cosmogenic helium that is produced by spallation reactions has a ^{3}He/^{4}He ratio of $\sim 0.2 - 0.3$ (Wieler 2002). This is 10^4 to 10^7 times higher than that of any terrestrial reservoir. This extreme enrichment of ^{3}He relative to ^{4}He in cosmogenic helium is the basis for its resolution from potentially interfering terrestrial components. For Earth-science applications, cosmogenic helium can be regarded as pure ^{3}He, since cosmogenic ^{4}He cannot be resolved against the natural background concentration of any geological material.

Inherited helium

To discriminate inherited ^{3}He trapped in fluid- and melt-inclusions from cosmogenic ^{3}He in the mineral lattice, two aliquots of a sample are usually analysed. Helium is extracted from one aliquot by *in vacuo* crushing. The other aliquot is melted to release the bulk of the helium (inclusion and matrix-hosted). Assuming that there is no contribution from radiogenic ^{4}He, the concentration of cosmogenic ^{3}He (^{3}He$_{cos}$) can be calculated (Craig and Poreda 1986, Kurz 1986b, a). This requires accurate measurement of the ^{3}He/^{4}He-ratio of the gas released by crushing (^{3}He/^{4}He)$_{crush}$, and melting (^{3}He/^{4}He)$_{melt}$, and the ^{4}He concentration determined from the melt extraction ^{4}He$_{melt}$:

$$^3\mathrm{He_{cos}} = \left[\left(\frac{^3\mathrm{He}}{^4\mathrm{He}} \right)_{melt} - \left(\frac{^3\mathrm{He}}{^4\mathrm{He}} \right)_{crush} \right] {}^4\mathrm{He_{melt}}$$

Prolonged and/or intense crushing of the minerals analysed can release matrix helium from the finest grains (Hilton *et al.* 1993, Scarsi 2000, Blard *et al.* 2006), probably by thermal diffusion (Blard *et al.* 2008b). If this effect is ignored it can lead to a calculated $^3He_{cos}$ value that is too low (Blard *et al.* 2006). Brief crushing, and crushing techniques that do not produce fine powder, or keep the sample cool, avoid this problem (Hilton *et al.* 1993, Scarsi 2000, Stuart *et al.* 2003, Blard *et al.* 2008b).

In cases where the assumption of no contribution by radiogenic ^4He is violated, e.g. in geologically old rocks, the calculated $^3He_{cos}$ will be significantly lower than the actual value (in the context discussed here, old signifies >50–100 ka in the case of most basalts). This may be corrected for via the determination of U and Th concentrations of the mineral and the surrounding rock (Dunai and Wijbrans 2000, Blard and Farley 2008). Chemical etching or physical abrasion can remove the radiogenic ^4He implanted from the surrounding rock matrix (Blard and Farley 2008), which is commonly enriched in U and Th relative to olivine or pyroxenes (Dunai and Wijbrans 2000, Blard and Farley 2008).

An alternative approach for geologically old rocks is to avoid mineral phases with trapped helium, e.g. by targeting microphenochryst phases in lava. This approach reduces the three-component problem to a two-component mixing between a radiogenic and a cosmogenic component, which can be solved with the two isotopes available (Williams *et al.* 2005). For this approach, it is important to characterize the $^3He/^4He$ ratio of the radiogenic end-member, which can be addressed by analysing shielded samples or by an analysis of equivalent material shielded from cosmic rays (Margerison *et al.* 2004). Alternatively it can be achieved by calculating the isotopic radiogenic and nucleogenic production (see below) in the rock (Margerison *et al.* 2004, Williams *et al.* 2005).

Production pathways and interferences

As cosmogenic ^3He is predominantly produced by spallation reactions, it can be produced on all elements present in a mineral, bar hydrogen (Section 1.5). Spallation production rates of ^3He in silicates are generally in the order of $\sim$100–120 atoms g^{-1} a^{-1} at sea level and high latitude (SLHL) (e.g. Licciardi *et al.* 1999, Dunai 2001a, Farley *et al.* 2006) and references therein). There is currently no consensus on the precise nature of the influence of major element composition on spallation production rates, however, variability of production rates in commonly used minerals, such as olivine and pyroxene, appear to be $<$10% (Schäfer *et al.* 1999, Masarik 2002, Kober *et al.* 2005, Farley *et al.* 2006). Muons have no resolvable contribution to cosmogenic ^{3}He production (Kurz 1986b, Sarda *et al.* 1992, Farley *et al.* 2006). A notable exception to the predominance of spallation reactions occurs in Li-rich minerals, for which a specific thermal-neutron capture reaction becomes important.

^{3}He is produced by a thermal-neutron capture reaction: ^{6}Li(n,α)^{3}H (β^-)^{3}He (δ = 955 barn). In the end both one α-particle and one ^{3}He are produced. Acquisition of two electrons by the α-particle results in ^{4}He. Consequently the ^{3}He/^{4}He ratio of nucleogenic helium is one. Neutrons for this reaction originate as thermalized neutrons of the cosmogenic cascade and thermalized neutrons from (α,n) reactions in rocks. The production of ^{3}He from cosmogenic thermal neutrons is significant in minerals with moderately high Li-concentrations (>5 ppm); and can account for 5–50% of cosmogenic production in typical accessory minerals in granite (Dunai *et al.* 2007). For minerals in basaltic lavas, the contribution of this reaction is lower, typically 1.5–6%, as a consequence of the generally low Li concentrations in mafic rocks (Dunai *et al.* 2007). The reaction products from the thermal neutrons produced by (α,n) reactions may accumulate over the geological age of a rock and can be a significant source of *nucleogenic* ^{3}He in minerals. The accumulation is a function of neutron flux in the rock, its age and thermal history and the Li-concentration of the mineral analysed (Andrews and Kay 1982, Andrews *et al.* 1989, Farley *et al.* 2006, Dunai *et al.* 2007, Blard and Farley 2008).

The term 'nucleogenic ^{3}He' customarily refers only to the ^{3}He produced by neutrons originating from (α,n) reactions in rocks. Until recently, thermalized cosmogenic neutrons (Dunai *et al.* 2007) were not considered as an important source for cosmogenic ^{3}He production.

Because the radiogenic end-member is important for exposure dating of geologically old rocks, the main controlling factors are briefly discussed here. Depending on the chemical environment, one in 10^6–10^8 α-particles produces a neutron by an (α,n)-reaction (Martel *et al.* 1990), $\geq$80% of which reach thermal energies (Morrison and Pine 1955, Andrews 1985, Liu *et al.* 1994a, Ballentine and Burnard 2002). Depending on the competition for capturing thermal neutrons by other elements in a rock, Li has a varying capacity to produce ^{3}He. For average crustal rocks, the expected radiogenic end-member ^{3}He/^{4}He ratio is 0.01–0.001 times the atmospheric ratio (cf. Ballentine and Burnard 2002). The ranges of α-particles (10s of microns) and neutrons (10s of centimetres) are different by a wide margin, and Li and the α-emitting elements U and Th do not necessarily reside in the same minerals. Therefore minerals within the same rock can have widely varying nucleogenic/radiogenic ^{3}He/^{4}He-ratios (Farley *et al.* 2006, Dunai *et al.* 2007).

Ejection distances

The significant ejection distances of radiogenic ^{4}He, nucleogenic ^{3}He and cosmogenic ^{3}He (and its precursor tritium), need consideration because they impinge on the corrections for radiogenic (Dunai and Wijbrans 2000, Farley *et al.* 2006) and nucleogenic (Farley *et al.* 2006, Dunai *et al.* 2007) helium, as well as on production-rate determinations (Farley *et al.* 2006, Dunai *et al.* 2007, Blard and Farley 2008). Generally, ejection distances of atomic nuclei are a function of their energy and their charge, and the composition and density of the material they have to traverse (Ziegler *et al.* 2008). The density of a mineral is a good guide for the relative stopping ranges in common minerals (Farley *et al.* 1996). Low-charge nuclei will travel further than higher-charge nuclei of the same kinetic energy, a relationship that is highly non-linear (Ziegler *et al.* 2008).

Radiogenic ^{4}He is emitted at energies between ~4 and 8.7 MeV, and has a range of between 11 and 34 microns in dense minerals such as apatite, titanite, and zircon (Farley *et al.* 1996). The ^{6}Li(n,α)^{3}H reaction ejects the ^{3}H nucleus with a kinetic energy of 2.7 MeV. This corresponds to a stopping range of about 30 µm in the same minerals (Farley *et al.* 2006, Dunai *et al.* 2007, Ziegler *et al.* 2008). These ejection distances (Fig. 2.2) largely control the radiogenic and nucleogenic helium budget of small-grained accessory minerals of dimensions similar to the ejection distances (Farley *et al.* 1996, Farley *et al.* 2006), but will also affect the rims of larger crystals (Dunai and Wijbrans 2000, Dunai *et al.* 2007).

Cosmogenic ^{3}He, and its precursor ^{3}H, are produced by spallation reactions that do not have strict upper energy limits; consequently, they may have energies of tens to hundreds of MeV. A 10 MeV ^{3}He nucleus will travel 54 microns in olivine (Fo$_{80}$) (Ziegler *et al.* 2008), which has a similar density to apatite. At 100 MeV this distance is 3 mm. Having a lower nuclear charge, cosmogenic tritium can travel much further than ^{3}He of the same energy: at 10 MeV and 100 MeV, a ^{3}H nucleus will travel 220 microns and 12 mm in olivine, respectively (Ziegler *et al.* 2008), i.e. four times further than ^{3}He (^{3}H is a precursor to ^{3}He, see above). The median kinetic energies of ^{3}H and ^{3}He from cosmic inter-actions are approximately 5–10 MeV and 10–20 MeV, respectively (Powell *et al.* 1959). Corresponding median ejection ranges for ^{3}H and ^{3}He in olivine are 70–220 microns and 54–170 microns, respectively

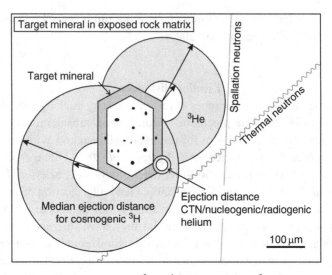

Fig. 2.2. Ejection distances of ^{3}He, ^{4}He and tritium (^{3}H) from nuclear reactions in surface rocks. An-inclusion bearing mineral grain is shown to illustrate effect of ejection distance on a moderately sized target mineral. The inclusions can be the host of magmatic noble gases, which may interfere with cosmogenic noble gas measurements (see text). Cosmogenic thermal neutron produced (CTN), nucleogenic and radiogenic helium has ejection distances in the order of 20–30 micron, depending on the density of the material. Median ejection ranges for spallogenic ^{3}H and ^{3}He in olivine are 70–220 micron and 54–170 micron, respectively. See main text for data sources and discussion.

(Ziegler *et al.* 2008). Median energies are a function of altitude and latitude (Powell *et al.* 1959).

These ejection distances are similar to the grain sizes of minerals used for exposure dating utilizing ^{3}He (Fig. 2.2). Consequently a considerable portion of ^{3}He measured in a mineral may have been produced outside that mineral, and have been implanted as ^{3}H or ^{3}He; likewise most of the ^{3}H and ^{3}He produced in a mineral may have come to rest in the surrounding host rock. Accounting for gains by implantation and losses by ejection requires accurate knowledge of the energy spectrum of particles emitted and of the spallogenic ^{3}H/^{3}He-production ratio (also called *branching ratio*, a term used for nuclear decay reactions that can proceed in two or more ways, and which denotes the ratio abundance of two alternative reaction products). The energy spectrum is not well known, and there is currently no consensus on the ^{3}H/^{3}He-branching ratio, estimates range between 1 and ~0.5 (Masarik 2002, Kober *et al.* 2005). If the production-rate contrast between host rock and minerals analysed is small, the relative effects on apparent mineral production rates will be small (Farley *et al.* 2006). Such a low contrast is probably realized for apatites in granite (Farley *et al.* 2006) and olivines and pyroxenes in basalt. Large contrasts can be expected, for example, for zircons in granites (Farley *et al.* 2006) or iron-oxides in acidic/felsic rocks (Dunai *et al.* 2007). In the latter cases apparent mineral production rates are a sensitive function of their grain size (Farley *et al.* 2006, Dunai *et al.* 2007).

Retentivity of minerals

Helium has the smallest van der Waals radius of all elements, 1.07 pm (Badenhoop and Weinhold 1997), and does not engage in chemical bonds in the natural environment. It can therefore diffuse relatively easily, and without any chemical interaction, using structural channels or structural defects (Trull *et al.* 1991, McDougall and Harrison 1999, Farley 2007). Consequently many minerals, including quartz and feldspars, cannot retain helium quantitatively at environmental temperatures (Trull *et al.* 1991, Shuster and Farley 2005, Farley 2007). At Earth-surface conditions, olivine (Trull *et al.* 1991), clinopyroxene, amphibole (Lippolt and Weigel 1988), garnet (Dunai and Roselieb 1996), zircon (Reiners and Farley 2000), titanite (Reiners and Farley 1999), magnetite (Hiyagon 1994) and haematite (Bähr *et al.* 1994, Wernicke and Lippolt 1994) quantitatively retain helium. Most of these minerals have been tested for cosmogenic studies e.g. (Trull *et al.* 1991, Schäfer *et al.* 1999, Margerison *et al.* 2004, Kober *et al.* 2005, Farley *et al.* 2006).

Applications

To date, the overwhelming majority of cosmogenic ^{3}He studies have exploited volcanic olivine and pyroxene (Niedermann 2002). Particularly for olivine, production rates have been extensively determined empirically (Cerling and Craig 1994a, Licciardi *et al.* 1999, Dunai and Wijbrans 2000, Dunai 2001a, Balco *et al.* 2008). As understanding of the production systematics of ^{3}He improves; particularly the dependence production rates on chemical composition, ejection distances of cosmogenic ^{3}H and ^{3}He, and the ^{3}H/^{3}He-branching ratio, the other helium retentive minerals will likewise become important cosmogenic tools.

A notable strength of cosmogenic ^{3}He lies in the relative ease of dating of young, phenocryst-bearing basalts (several thousand years; e.g. Cerling and Craig 1994; Liccardi *et al.* 1999), which often cannot be dated reliably with other techniques or only with considerable effort. Furthermore, ^{3}He

is also important for the study of very old surfaces that cannot be dated
with commonly used radionuclides (e.g. Schäfer *et al.* 1999, Magerison
et al. 2004, Evenstar *et al.* 2009).

2.2.2 Cosmogenic neon (^{21}Ne, ^{22}Ne)

Neon is slightly more abundant than helium in the atmosphere, with
18.2 ppm Ne (by volume) in air (Eberhardt *et al.* 1965, Porcelli *et al.*
2002). It has three stable isotopes ^{20}Ne, ^{21}Ne and ^{22}Ne, of which ^{20}Ne is
the most abundant (90.50% of atmospheric neon), the atmospheric
^{21}Ne/^{20}Ne- and ^{22}Ne/^{20}Ne-ratios are 0.00296 and 10.20, respectively
(Eberhardt *et al.* 1965). All three neon isotopes are produced at similar
rates by spallation reactions in rocks (Niedermann 2002). Due to the
lower natural abundance, the ^{21}Ne and ^{22}Ne budgets in samples are more
noticeably affected by cosmogenic production than that of ^{20}Ne. Generally it is the concentration of cosmogenic ^{21}Ne that is used to calculate
exposure ages and/or erosion rates.

Inherited neon

Atmospheric and cosmogenic neon are present in all exposed samples. In
addition, mantle and crustal neon may be trapped in fluid and melt
inclusions. Further, nuclear reactions involving oxygen, fluorine or magnesium can produce neon isotopes in minerals over the lifetime of rocks.
The identification of and correction for the interfering non-cosmogenic
components in a sample is largely achieved using a neon three-isotope
diagram (Fig. 2.3) (Niedermann *et al.* 1994, Niedermann 2002). In such a
diagram, mixtures between two components lie on straight lines linking
the end-members. The mixing line between atmospheric and cosmogenic
Ne is well defined (see below), such that significant additional components can normally be resolved with confidence.

Mantle-derived neon is enriched in ^{20}Ne and ^{21}Ne relative to air
(cf. Graham 2002). In mantle-derived material that may be collected for
exposure dating (typically phenocrysts in subaerially exposed basalts) the
mantle components can often not be resolved next to the much more
abundant atmospheric neon adsorbed on or trapped in the sample. Thus,
mantle neon is *usually* not a concern for cosmogenic applications. If
mantle neon is present in significant amounts, it can be clearly identified
in the neon three-isotope diagram. The mantle trend, along which measurements would be displaced, is almost perpendicular to the spallation
line (Fig. 2.3).

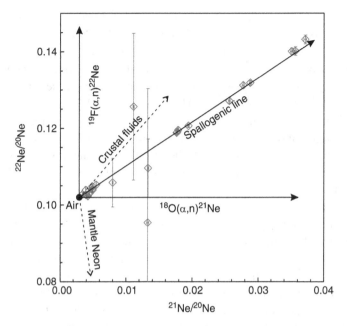

Fig. 2.3. Neon three-isotope diagram. Mixtures of two neon components lie on lines in such a diagram (Niedermann *et al.* 1994, Niedermann 2002). The spallogenic line indicates the mixture of air and the spallogenic end-member, which are present in all surface samples. Additional components may be nucleogenic ^{22}Ne and ^{21}Ne produced by (α,n) reactions on fluorine and oxygen, respectively; and mantle (Graham 2002) and crustal neon (Kennedy *et al.* 1990). The latter may be trapped in fluid and melt inclusions. Underlain in grey are Ne-data from Dunai *et al.* (2005) to show a case where all data lie within uncertainties (i.e. $\pm 2\sigma$; for clarity $\pm 1\sigma$ is shown) on the spallation line, and that ^{21}Ne/^{20}Ne ratios usually have smaller uncertainties than ^{22}Ne/^{20}Ne. This is the main reason why age-information is usually obtained from the ^{21}Ne/^{20}Ne-ratio and corresponding non-atmospheric ^{21}Ne-concentration (Hetzel *et al.* 2002, Niedermann 2002).

Crustal fluids trapped in minerals are typically enriched in ^{21}Ne and ^{22}Ne. This enrichment arises from the time-integrated production by ^{18}O$(\alpha,n)^{21}$Ne and ^{19}F$(\alpha,n)^{22}$Ne$(\beta^{-})^{22}$Ne reactions in the source region of the fluids. The isotopic composition depends on the O/F ratio in the source region, and differential release of ^{21}Ne and ^{22}Ne from minerals (Kennedy *et al.* 1990, Ballentine and Burnard 2002). ^{22}Ne is more abundant in crustal fluids than would be expected simply from the average O/F ratio in rocks. This is because ^{22}Ne is preferentially released from F-bearing micas. ^{21}Ne is also produced in micas, however, as well as in more retentive tectosilicates and oxides. Therefore, ^{21}Ne is, on average, less easily released from minerals (Kennedy *et al.* 1990, Ballentine and Burnard 2002). The characteristics of the crustal fluid source and, there-fore, the neon isotopic composition of crustal fluids trapped in minerals, is in most cases unknown *a priori*. The latter can be determined by *in vacuo* crushing experiments (Hetzel *et al.* 2002a) or by analysing equivalent shielded samples. The ability to characterize and correct for the crustal component is important, because crustal neon may mimic a cosmogenic signal and/or affect the position of the non-cosmogenic mixing end-member in the neon three-isotope plot (Hetzel *et al.* 2002a, Niedermann 2002).

The nuclear reactions can also produce nucleogenic neon in the lattice of the minerals analysed. This is released together with the cosmogenic neon during heating of the sample. Quartz and olivine, the two minerals mostly used in cosmogenic neon studies, generally have low uranium and thorium concentrations. However, U and Th in mineral inclusions or neighbouring minerals can be a source of α-particles. The limited range of α-particles (tens of microns), means that the removal of the rims of mineral grains, e.g. by acid etching or physical abrasion, can reduce this nucleogenic component.

Stepwise heating experiments are usually used to establish whether the neon isotopic composition is a two/three-component mixture between non-cosmogenic (atmospheric, nucleogenic) and cosmogenic end-members. If all extraction steps of a sample have isotopic compositions commensurate with simple two-component mixing, i.e. lie on the spalla-tion line (Fig 2.3), or three-component mixing on a trend parallel to it (Hetzel *et al.* 2002a, Niedermann 2002), the cosmogenic ^{21}Ne concen-tration can usually be calculated with confidence (Niedermann 2002).

Production pathways and interferences

Cosmogenic neon is produced by high-energy and fast neutrons (see Section 1.1). Recent findings indicate that muons contribute ~3.6% to ^{21}Ne production in quartz at SLHL (Balco and Shuster 2009). Earlier measurements did not find a resolvable contribution of muons to cosmo-genic Ne production (Sarda *et al.* 1992, Farley *et al.* 2006). Published values for the production rate of ^{21}Ne in quartz are between 18.4 and 20 atoms g^{-1} a^{-1}, at SLHL (Niedermann 2000, Balco and Shuster 2009, Goethals *et al.* 2009a).

Target elements with masses similar to neon, i.e. Na, Mg, Al and Si, have the largest reaction cross-sections (Schäfer *et al.* 1999, Masarik 2002,

Kober *et al.* 2005). The (n,α) reactions on ^{24,25}Mg producing ^{21,22}Ne are energetically favoured and can be accomplished by fast neutrons (Fig. 1.7); the production from other elements is mostly achieved by high-energy neutrons. Consequently, the production rate of ^{21}Ne from Mg is, for example, $\sim$4.2–4.4 times higher than from Si. Accordingly there is a strong compositional dependency of cosmogenic neon production rates (Schäfer *et al.* 1999, Masarik 2002, Niedermann 2002, Kober *et al.* 2005). The element-specific production-rate estimates of Masarik (2002), Schäfer *et al.* (1999) and Kober *et al.* (2005) agree well for Si (3% variance), Al (7% variance) and Mg (6% variance). However, large differences (>100%) exist in the evaluation of production rates from Na and Ca (Masarik 2002, Kober *et al.* 2005). The latter may merit attention when using Na- and Ca-bearing minerals (e.g. some members of the pyroxene, amphibole and feldspars group of minerals) for exposure dating utilizing cosmogenic neon. Due to the isotope-specific elemental reaction cross-sections, the slope s of the spallation line (Fig. 2.3) depends on the mineral composition (Niedermann 2002). For quartz and augite (a pyroxene) this slope has been determined experimentally. The reported values of s for quartz are 1.120 ± 0.020 (Niedermann *et al.* 1993) and 1.143 ± 0.038 (Schäfer *et al.* 1999), hence in excellent agreement. Equally, the reported values for augite of 1.055 ± 0.017 (Bruno *et al.* 1997) and 1.069 ± 0.035 (Schäfer *et al.* 1999) agree unambiguously.

Retentivity of minerals

Similarly to helium, neon does not engage in chemical bonds in the natural environment. Neon has, however, a 14% larger van der Waals radius than helium (Badenhoop and Weinhold 1997) and is therefore better retained in minerals. Tectosilicates such as quartz and sanidine quantitatively retain cosmogenic neon under a wide range of environmental conditions (Kober *et al.* 2005, Shuster and Farley 2005), whereas helium is not reliably retained in these minerals. However, terrestrial feldspars other than sanidine do not retain neon quantitatively (Bruno *et al.* 1997, Schäfer *et al.* 1999, Kober *et al.* 2005). All helium-retentive minerals, including olivine and pyroxenes (see Sect. 2.2.1), retain cosmogenic neon.

Applications

To date, the overwhelming majority of cosmogenic neon studies rely on quartz, olivine and pyroxene, with quartz taking the lion's share (Niedermann 2002). The production rates in quartz are well established and reliably cross-calibrated with other radionuclides (Niedermann 2000, Balco and Shuster 2009, Goethals *et al.* 2009a). For olivine and other Ca- and Na-poor silicates, production rates are reasonably well constrained (Niedermann 2002). As our understanding of the dependence of production rates of cosmogenic neon on chemical composition improves, more minerals will become useful tools for cosmogenic applications.

The particular strength of cosmogenic neon lies in the applicability to quartz on very old surfaces (>5 Ma) or very slowly eroding surfaces (<10 cm Ma^{-1}) where it can provide unique information that is unattainable with most cosmogenic radionuclides (Bruno *et al.* 1997, Schäfer *et al.* 1999, Van der Wateren *et al.* 1999, Dunai *et al.* 2005, Schäfer *et al.* 2006). Further, the small sample sizes that are necessary for neon analysis (commonly ~100 mg) permit single-clast sediment-provenance studies (Codilean *et al.* 2008), which are difficult to achieve with other nuclides. The combination of ^{21}Ne with ^{10}Be or ^{26}Al provides the prospect of resolving exposure/erosion histories of old and/or slowly eroding surfaces (500 ka–5 Ma; 5–50 cm Ma^{-1}; Kober *et al.* 2007).

2.2.3 Cosmogenic argon (^{36}Ar, ^{38}Ar)

Argon is the most abundant noble gas in the Earth's atmosphere: there is 0.93% argon (by volume) in air (Nier 1950, Porcelli *et al.* 2002). It has three stable isotopes ^{36}Ar, ^{38}Ar and ^{40}Ar, of which ^{40}Ar is the most abundant (99.6% of atmospheric argon). The atmospheric ^{40}Ar/^{36}Ar and ^{38}Ar/^{36}Ar ratios are 295.5 and 0.1880, respectively (Nier 1950, Porcelli *et al.* 2002). Due to their lower natural abundance, the ^{36}Ar and ^{38}Ar budgets in minerals may be noticeably affected by cosmogenic production (Renne *et al.* 2001). However, even the rarest Ar isotope, ^{38}Ar, has an abundance of 5.9 ppm (by volume) in air, which is $8 \cdot 10^7$ and 121 times more abundant than ^{3}He and ^{21}Ne, respectively. Detection of cosmogenic ^{38}Ar and ^{36}Ar is therefore complicated by a comparatively high background concentration, which may be aggravated by the increasing ability of heavier noble gases to adsorb to mineral surfaces (Ozima and Podosek 2001).

Due to the decay of ^{40}K to ^{40}Ar (the basis of K–Ar dating) the ^{40}Ar/^{36}Ar ratio of argon trapped or produced in minerals is variable and *a priori* unknown. Hence investigations of ^{38}Ar/^{40}Ar or ^{36}Ar/^{40}Ar ratios can not be used to assess cosmogenic production of ^{38}Ar or ^{36}Ar (i.e. in a manner comparable to the neon three-isotope plot, Fig. 2.3). The ^{38}Ar/^{36}Ar ratio of essentially all terrestrial material unaffected by cosmogenic production is uniform and identical to the atmospheric composition (Renne *et al.* 2001). Consequently cosmogenic argon can be detected by changes relative to this atmospheric ratio. This process can be complicated by different production mechanisms and/or target elements producing cosmogenic ^{38}Ar and ^{36}Ar (see below).

Production pathways and interferences

Cosmogenic argon is mainly produced by spallation reactions on K, Ca and to a lesser extent from Ti and Fe. The SLHL production rate of ^{38}Ar from Ca is in the order of 200 atoms g^{-1} a^{-1} (Lal 1991, Niedermann *et al.* 2007); the ^{38}Ar/^{36}Ar production ratio from calcium is 1.5 ± 0.2 (Hohenberg *et al.* 1978, Niedermann *et al.* 2007). Production rates of ^{38}Ar and ^{36}Ar from K are both about 1.5 times higher than that of ^{38}Ar from Ca, i.e. about 300 atoms ^{38}Ar and ^{36}Ar g^{-1} a^{-1} (Hohenberg *et al.* 1978, Niedermann *et al.* 2007). The production rates of ^{38}Ar and ^{36}Ar from Ti and Fe are ~6% and 0.3–0.4% of the corresponding Ca production rate, respectively (Hohenberg *et al.* 1978).

Muons produce ^{38}Ar directly from K and Ca (reactions: ^{39}K$(\mu^-,n)^{38}$Ar and ^{40}Ca$(\mu^-,pn)^{38}$Ar); and ^{36}Ar indirectly via the precursor nuclide ^{36}Cl (see below and Section 2.3.4; reactions: ^{40}Ca$(\mu^-,\alpha)^{36}$Cl$(\beta^-)^{36}$Ar, ^{39}K$(\mu^-,2\,np)^{36}$Cl$(\beta^-)^{36}$Ar).

An important detail in the ^{36}Ar reaction pathway is the production of ^{36}Ar via ^{36}Cl decay $(T_{1/2} = 301\,ka)$ that occurs concurrently with direct production of ^{36}Ar. ^{36}Cl is in itself a cosmogenic nuclide, produced by spallation and muons on the same target elements as cosmogenic argon, and by neutron capture on ^{35}Cl (see 2.3.4). This pathway via the relatively long-lived precursor ^{36}Cl has as consequence that the spallation production ratio of ^{38}Ar/^{36}Ar from Ca is 3.1 for very recent exposure, and 1.5 for long exposure, when ^{36}Cl reached secular equilibrium (>1 Ma; calculated using values of Hohenberg *et al.* 1978; after Renne *et al.* 2001). For young (<1 Ma) and (usually) unknown exposure histories the ^{38}Ar/^{36}Ar production ratio is therefore not well constrained. The relative and absolute production rates for ^{36}Ar given above are only valid for old surfaces (>1 Ma).

Thermal-neutron capture by ^{35}Cl and subsequent β-decay of ^{36}Cl is a significant production pathway for cosmogenic ^{36}Ar in chlorine-bearing minerals. For instance, a mineral with 100 ppm Cl, in a granite or basalt matrix, produces ~3 atoms ^{36}Cl g^{-1} a^{-1} at SLHL; Phillips *et al.* 2001. The corresponding production pathway of ^{38}Ar via ^{37}Cl has a much lower neutron-capture cross-section (0.4 barn as compared to 44 barn for ^{35}Cl; ^{35}Cl is also more abundant), and can be neglected. The thermal-neutron flux in a rock is a sensitive function of the water content of the rock and the environment, as well as the erosion rate (Phillips *et al.* 2001). Because these parameters can vary during the required long exposures (>1 Ma, see above) the accurate determination of the $^{38}Ar/^{36}Ar$ production rate, a prerequisite to determining cosmogenic ^{38}Ar concentrations, is difficult in chlorine-rich minerals (Niedermann *et al.* 2007).

Retentivity of minerals

Argon has a much larger van der Waals radius than helium or neon (by 66% and 46%, respectively; Badenhoop and Weinhold 1997) and is much better retained in minerals. Most silicates and all feldspars can quantitatively retain argon under environmental conditions (Lippolt and Weigel 1988, Fortier and Giletti 1989, McDougall and Harrison 1999). Helium- or neon-retentive minerals generally retain argon (Lippolt and Weigel 1988).

Application

The considerable atmospheric background, and the 'branched' ^{36}Ar production pathways (i.e. direct and via ^{36}Cl), will probably limit the application of cosmogenic argon to environments with old surfaces and low erosion rates. Feldspars are often abundant in rocks devoid of olivine, pyroxene or quartz that are routinely used in conjunction with the other stable cosmogenic nuclides (i.e. ^{3}He and ^{21}Ne, respectively; see Sections 2.2.1 and 2.2.2). In these situations the application of cosmogenic ^{38}Ar may present itself as a unique tool.

2.2.4 Cosmogenic krypton and xenon

Krypton and xenon are both rare noble gases with concentrations of 1.1 and 0.09 ppm (by volume) in air, respectively (Porcelli *et al.* 2002). Krypton has six stable isotopes (^{78}Kr, ^{80}Kr, ^{82}Kr, ^{83}Kr, ^{84}Kr, ^{86}Kr), and xenon, nine (^{124}Xe, ^{126}Xe, ^{128}Xe, ^{129}Xe, ^{130}Xe, ^{131}Xe, ^{132}Xe, ^{134}Xe, ^{136}Xe) (Porcelli *et al.* 2002). The lightest isotopes ^{78}Kr, ^{80}Kr and ^{124}Xe, ^{126}Xe, ^{128}Xe have the lowest relative abundances (all < 2.5%; Porcelli *et al.* 2002). The long-lived radionuclide ^{81}Kr ($T_{1/2} = 229 \pm 11$ ka; Baglin 2008) is used for exposure dating of extraterrestrial material (Eugster 1988).

Production pathways and interferences

Cosmogenic Kr can be produced by spallation from Rb, Sr, Y and Zr and heavier elements, as well as by neutron capture by Br (^{80}Kr, ^{82}Kr). Cosmogenic Xe can be produced by spallation from Cs, Ba, La, Ce and all heavier rare earth elements (REE), as well as by neutron capture via I (^{128}Xe) and Te (^{129}Xe, ^{131}Xe) (Browne and Berman 1973, Srinivasan 1976, Bernatorwicz *et al.* 1993).

Spontaneous fission of ^{238}U is the dominant mechanism for the production of ^{83,84,86}Kr and ^{129,131,132,134,136}Xe in crustal rocks (cf. Ballentine and Burnard 2002). Spontaneous fission of ^{232}Th and induced fission of ^{232}Th and ^{235}U also contribute to the production of these isotopes (cf. Ballentine and Burnard 2002). The ejection distance of fission products is of the order of 10 microns (Guedes *et al.* 2007), such that similar considerations as discussed for helium (Section 2.2.1) pertaining to ejection losses or implantation gains have to be made for fine-grained material (Pinti *et al.* 2001). The fissiogenic production of ^{80,82}Kr and ^{124,126,128,130}Xe is negligible (Ballentine and Burnard 2002) such that it should be possible to resolve cosmogenic production of these isotopes in U- and Th-bearing minerals.

Similar to the other cosmogenic radionuclides (Section 2.3), the non-cosmogenic background concentration of ^{81}Kr ($T_{1/2}$ = 229 ± 11 ka; Baglin 2008) is reduced by radioactive decay.

Retentivity of minerals

Retentivity of minerals for krypton and xenon is high. Besides their large van der Waals radii, which stand in the way of fast diffusion, they adsorb readily on mineral surfaces due to their large, polarizable electron shells (Ozima and Podosek 2001).

Applications

To date, there is only one study published on cosmogenic xenon (in barite, BaSO$_4$; Srinivasan 1976), and no application of terrestrial cosmogenic krypton.

Conceivable target minerals for cosmogenic krypton could be Sr-bearing carbonates or celestine (SrSO$_4$), and potentially zircon (ZrSiO$_4$) or baddeleyite (ZrO$_2$). For cosmogenic xenon, feasible targets are barite (Srinivasan 1976), and monazite or xenotime ((La,Ce,REE)PO$_4$). These minerals are reasonably common, some slightly exotic (e.g. baddeleyite), but may be useful in specialist applications. However, essentially all groundwork, ranging from element-specific production rates to the resolution of interferences, remains to be covered before cosmogenic krypton or xenon can be used as a tool for Earth-surface sciences.

2.3 Cosmogenic radionuclides

Unlike the stable cosmogenic nuclides that all belong to the noble gases, the useful cosmogenic radionuclides have members in five different element categories: alkali earth metals (^{10}Be, ^{41}Ca), transition elements (^{53}Mn), halogens (^{36}Cl), other metals (^{26}Al) and non-metals (^{14}C). Their low natural background concentration in minerals or rocks relies on the depletion of any inherited nuclides by radioactive decay rather than on favourable physical or chemical fractionation. Cosmogenic radionuclides have half-lives that are typically much shorter than the geological age of the sample material analysed, and after 5–6 half-lives, a radionuclide can be practically considered extinct (Faure and Mensing 2004). The half-lives are, in increasing order: ^{14}C, 5730 a; ^{41}Ca, 104 ka; ^{36}Cl, 301 ka; ^{26}Al, 708 ka; ^{10}Be, 1.36 Ma and ^{53}Mn, 3.7 Ma (^{14}C: Lederer *et al.* 1978; ^{26}Al: Nishiizumi 2004; ^{10}Be: Nishiizumi *et al.* 2007; ^{36}Cl Holden 1990; ^{41}Ca: Kutschera *et al.* 1992; ^{53}Mn: Honda and Imamura 1971).

Some cosmogenic radionuclides (^{10}Be, ^{14}C, ^{36}Cl) are produced at significant rates in the atmosphere and can be adsorbed onto or otherwise incorporated into sample material. For these isotopes, usually a considerable chemical preparation effort is aimed at removing this 'atmospheric' or 'meteoric' component.

2.3.1 Cosmogenic beryllium (^{10}Be)

Beryllium has one stable nuclide (^{9}Be) and two cosmogenic radionuclides, ^{7}Be and ^{10}Be; ^{7}Be has too short a half-life ($\sim$53 days) to be useful for *in situ* applications; the half-life of ^{10}Be is 1.36 ± 0.07 Ma (Nishiizumi *et al.* 2007). At the time of writing, however, this value is not yet universally adopted. The alternative value used is 1.51 ± 0.06 Ma (Hofmann *et al.* 1987). In practice, if half-lives are consistently used in the data reduction (e.g. Balco *et al.* 2008), the choice of half-life has no effect on dating young samples (of the order of 10^4 years), for which decay during exposure is not significant. However, for samples where ^{10}Be decay affects the total ^{10}Be inventory, e.g. burial dating (Chapter 4) or samples near equilibrium (in the order of 10^6 years; Section 2.1.2) the choice of half-life does affect the interpretation (Nishiizumi *et al.* 2007, Balco *et al.*

2008). At the time of writing, two additional independent evaluations of the ^{10}Be half-life have been concluded; both report a value of 1.39 Ma, with a stated uncertainty of about 1% (Chemeleff *et al.* 2009, Korschinek *et al.* 2009). These new values are in agreement with the revised value of Nishiizumi *et al.* (2007), and should allow the formulation of an accurate consensus on the ^{10}Be half-life for use in exposure dating and burial dating.

Meteoric/Atmospheric ^{10}Be

Cosmogenic ^{10}Be is produced in the atmosphere by spallation reactions on nitrogen and oxygen. This production is of the order of 10^3 times faster than the average rate in rocks (Gosse and Phillips 2001). Atmospheric ^{10}Be eventually precipitates and can be adsorbed by surface materials (the adsorbed meteoric ^{10}Be is also referred to as 'garden variety' ^{10}Be, e.g. Nishiizumi *et al.* 1986). In the case of quartz, sequential chemical dissolution can reliably remove the meteoric component (Kohl and Nishiizumi 1992). Such a pretreatment is a prerequisite for all terrestrial *in situ* applications of ^{10}Be. The difficulty of removing meteoric ^{10}Be from minerals other than quartz stands in the way of the widespread use of other mineral phases for *in situ* ^{10}Be applications. However, successful procedures have been reported for olivine (Nishiizumi *et al.* 1990, Seidl *et al.* 1997, Blard *et al.* 2008a), pyroxene (Blard *et al.* 2008a), sanidine (Kober *et al.* 2005) and carbonates (Braucher *et al.* 2005, Merchel *et al.* 2008b). This demonstrates that mineral-specific solutions to the problem of meteoric ^{10}Be can often, but not always (Merchel *et al.* 2008b), be found.

Production pathways and interferences

In rocks, ^{10}Be is mainly produced by spallation reactions from O, and to a lesser extent from heavier elements like Mg, Al, Si and Ca (Masarik 2002, Kober *et al.* 2005). In carbonates, C is a significant target element for ^{10}Be (Braucher *et al.* 2005). At SLHL, stopped negative muon and fast muon interactions together account for 3.6% of ^{10}Be production in quartz (Heisinger *et al.* 2002a), a value that increases rapidly with depth below the surface (Section 1.5). The SLHL production rate of ^{10}Be in quartz is about 4.5 atoms g^{-1} a^{-1} (cf. Balco *et al.* 2008; recalculated for $T_{1/2} = 1.36$ Ma).

Interactions of low-energy α-particles with ^{7}Li can produce ^{10}Be via an (α,n) reaction (Sharma and Middleton 1989). However, because Li, U and Th are trace elements in all minerals that have been used for *in situ*

[10]Be applications to date, this reaction is not significant in practice (Brown *et al.* 1991).

In Be-bearing minerals [10]Be can be produced by thermal neutron capture of [9]Be (n,γ-reaction; δ = 7.8 mbarn, Nishiizumi *et al.* 2007; Sharma and Middleton, 1989). While this reaction is not directly relevant for *in situ* applications (because only Be-poor minerals are analysed), this reaction is limiting the minimum [10]Be/[9]Be ratios that can be achieved for Be-carriers that are derived from deeply shielded natural materials (Schäfer, personal communication). These carriers are required for [10]Be analysis (Section 2.4).

In conclusion, there are no non-cosmogenic [10]Be production pathways that need consideration for standard *in situ* applications.

Applications

[10]Be in quartz is the workhorse for *in situ* applications, and the vast majority of *in situ* studies use [10]Be, either alone or in conjunction with [26]Al, [21]Ne or [14]C (Sections 2.1 and 4.2). Quartz is a mineral that can be found in a wide range of geological settings. Due to the long half-life of [10]Be, and the increasingly low analytical backgrounds that can be realized, samples from settings covering the entire Quaternary, including historic times (Schaefer *et al.* 2009), can be analysed. Dating of Pliocene samples is possible, but becomes challenging for the Early Pliocene settings, when [10]Be approaches saturation.

In conjunction with [26]Al, [10]Be is used for burial dating sediments (Granger *et al.* 2001, Granger and Muzikar 2001; Section 4.2), and, in conjunction with [26]Al, [21]Ne or ^{14}C, it can provide information on complex exposure histories (Lal 1991, Nishiizumi *et al.* 1991a, Gosse and Phillips 2001, Kober *et al.* 2005, Miller *et al.* 2006; Section 4.1).

The development of minerals other than quartz, such as olivine, pyroxene, sanidine and carbonates, will allow the combination of [10]Be with other cosmogenic nuclides, such as [3]He and [36]Cl. This would allow addressing complex exposure histories in cases where this currently is not possible.

2.3.2 Cosmogenic carbon (^{14}C)

Carbon has two stable nuclides, ^{12}C and [13]C. The cosmogenic radionuclide ^{14}C has a half-life of 5730 ± 30 years (Lederer *et al.* 1978). Decay of cosmogenic ^{14}C that is produced in the atmosphere and incorporated into biological material is the basis of the well-established radiocarbon technique (Libby 1946, Libby *et al.* 1949). The development of *in situ* methodology took off more recently (Jull *et al.* 1994, Lal and Jull 1994, Handwerger *et al.* 1999, Lifton *et al.* 2001, Kim *et al.* 2007).

Meteoric/Atmospheric ^{14}C

Thermal neutrons of the cosmic-ray cascade produce abundant ^{14}C from ^{14}N in the atmosphere (via the reaction ^{14}N(n,p)^{14}C), where the ^{14}C produced equilibrates with CO_2 in the atmosphere (Libby *et al.* 1949). The atmospheric nuclear test explosions of the 1950s and 1960s significantly added to the ^{14}C inventory of the atmosphere and the biosphere (Vries 1958, Taylor and Berger 1967). Atmospheric CO_2 is incorporated into biological material by respiration, and it is washed out as carbonic acid in rainfall and can potentially contaminate *in situ* samples. Any non-fossil organic material (either particulate matter or organic molecules) adhering to samples used for *in situ* analysis needs to be quantitatively removed prior to analysis. This can be achieved by treatment with oxidizing acids and thermal combustion of organic matter (Lifton *et al.* 2001). Also, secondary carbonates produced by chemical weathering by carbonic acid need to be removed quantitatively. Due to the difficulty in achieving the latter in most silicate rocks (Gosse and Phillips 2001), development of the method is currently focused on quartz (Lifton *et al.* 2001, Kim *et al.* 2007). One study focused on carbonates (Handwerger *et al.* 1999). Careful sample handling in the laboratory and analytical procedures aim to reduce contamination by organic matter and atmospheric CO_2 at any processing stage (Lifton *et al.* 2001).

Production pathways and interferences

In rocks, ^{14}C is mainly produced by spallation reactions from O and to a lesser extent from Mg, Al and Si (Masarik 2002). At SLHL, stopped negative muons and fast muons account for 14% of ^{14}C production in quartz (Heisinger *et al.* 2002a), a value that increases rapidly with depth in the subsurface (Section 1.5). At SLHL the total production rate of cosmogenic ^{14}C in quartz is about 17 atoms g^{-1} a^{-1} (Dugan *et al.* 2008).

In cases where nitrogen is a major constituent of fluid inclusions in samples, the production of ^{14}C by thermalized cosmic-ray neutrons (Section 1.4) may be measurable (Kim *et al.* 2007).

Nucleogenic ^{14}C from the thermal-neutron flux generated in rocks (Section 1.5) is usually not significant. This is due to the commonly low concentration of the relevant elements (U, Th) in sample material (quartz, carbonate), the short integration time (short half-life of ^{14}C)

(Barker *et al*. 1985, Jull *et al*. 1987), and the predominance of cosmic-ray-derived thermal neutrons over neutrons derived from (α,n) reactions in surface rocks (Section 1.5).

Applications

Currently, quartz is the only target mineral that is quasi-routinely analysed (Lifton *et al*. 2001, Kim *et al*. 2007, Anderson *et al*. 2008, Dugan *et al*. 2008, Lifton 2008). Ice (Jull *et al*. 1994) and carbonates (Handwerger *et al*. 1999) have also been used as target materials.

Due to its short half-life, ^{14}C reaches saturation quickly (Section 2.1). This makes it uniquely suitable to study short-term erosion rates (Gosse and Phillips 2001; Section 4.3) and young (Holocene) burial histories (Miller *et al*. 2006, Anderson *et al*. 2008; Section 4.2).

2.3.3 Cosmogenic aluminium (^{26}Al)

Aluminium has one stable nuclide (^{27}Al); the cosmogenic radionuclide ^{26}Al has a half-life of 708 ± 17 ka (Nishiizumi 2004). As aluminium is a major element in many minerals, and an appreciable trace element in most, the background of stable ^{27}Al is an important consideration in ^{26}Al analysis. If samples contain too much ^{27}Al (several hundred ppm) the ^{26}Al/^{27}Al ratio becomes increasingly difficult to measure (Section 2.4). Consequently, only low-Al quartz is currently used for *in situ* applications.

Meteoric/Atmospheric ^{26}Al

Due to the lack of appropriate target elements in the atmosphere, production of atmospheric ^{26}Al is low. Consequently, meteoric ^{26}Al is usually not a major concern in the application of *in situ* ^{26}Al (Gosse and Phillips 2001). More often than not, ^{26}Al is analysed in conjunction with ^{10}Be. The rigorous cleaning procedure required for ^{10}Be (Kohl and Nishiizumi 1992) removes any meteoric ^{26}Al that may be present. Moreover, the sequential etching lowers the ^{27}Al-concentration in the quartz separate (Kohl and Nishiizumi 1992).

Production pathways and interferences

In typical rocks, ^{26}Al is mainly produced by spallation reactions from ^{27}Al and Si (Masarik 2002). In non-silicate minerals, production from P, S, Cl, K and Ca could become important. Protons from the cosmic-ray cascade can produce ^{26}Al via the ^{26}Mg(p,n)^{26}Al reaction (Skelton and Kavanagh 1987, Masarik 2002). At SLHL, stopped negative muons and fast muons together account for 4.5% of ^{26}Al production in quartz (Heisinger *et al.* 2002a), a value that increases rapidly with depth in the subsurface (Section 1.5). At SLHL the total production rate of cosmogenic ^{26}Al in quartz is about 30 atoms g^{-1} a^{-1} (cf. Balco *et al.* 2008).

In environments that are rich in U, Th and/or Na, appreciable amounts of ^{26}Al can be produced via the ^{23}Na(α,n)^{26}Al reaction (Sharma and Middleton 1989, Placzek *et al.* 2007). This non-cosmogenic reaction can be used to date evaporites (Placzek *et al.* 2007).

Summarizing, for standard applications of ^{26}Al, only the spallogenic and muogenic production from Si is relevant.

Applications

Currently, quartz is the only target mineral used for ^{26}Al. It is frequently applied in conjunction with ^{10}Be; few studies rely on ^{26}Al alone. This may be because the accuracy and reproducibility of ^{10}Be concentration determinations is often better than for ^{26}Al. Due to the significant background concentration in samples, the ^{27}Al content needs to be determined analytically. The analytical uncertainty of this determination has a direct effect on calculated ^{26}Al concentrations.

Together, the ^{10}Be $-^{26}$Al couple is regularly used to constrain complex exposure histories (Lal 1991, Nishiizumi *et al.* 1991a, Gosse and Phillips 2001; Section 4.1) and for burial dating of sediments (Granger *et al.* 2001, Granger and Muzikar 2001; Section 4.2).

2.3.4 Cosmogenic chlorine (^{36}Cl)

Natural chlorine has two stable nuclides (^{35}Cl and ^{37}Cl; with a ^{37}Cl/^{35}Cl ratio of 0.319); the cosmogenic radionuclide ^{36}Cl has a half-life of 301 $\pm$ 2 ka (Holden 1990).

Meteoric/Atmospheric ^{36}Cl

There is an appreciable production of ^{36}Cl in the atmosphere by spallation from Ar, which constitutes 0.9% of air (by volume). However, due to the hydrophilic nature of Cl, and the good solubility of most secondary chlorine-bearing minerals (Deer *et al.* 1992), it tends not to adhere to rocks or minerals. Grinding and rinsing in water is adequate to remove meteoric Cl from most silicates (Zreda *et al.* 1991). If secondary minerals are macroscopic (e.g. as carbonate crusts), meteoric ^{36}Cl trapped in fluid inclusions can be a significant source of contamination (Merchel, personal communication).

Production pathways and interferences

In typical rocks, ^{36}Cl is mainly produced by spallation reactions from K and Ca, and, to a lesser extent, from Ti and Fe (Stone *et al.* 1996d, Evans *et al.* 1997, Phillips *et al.* 2001, Swanson and Caffee 2001, Masarik 2002).

Capture of thermalized cosmogenic neutrons (Section 1.5) by ^{35}Cl can produce significant amounts of ^{36}Cl via the ^{35}Cl(n,γ)^{36}Cl reaction (Phillips *et al.* 2001). This may contribute significantly to the bulk of the cosmogenic ^{36}Cl in a sample (Schimmelpfennig *et al.* 2009). The relevant neutron flux, i.e. the production yield of this production pathway, is modulated by the water content of the rock (Phillips *et al.* 2001), as well as the moisture content of the environment (Desilets *et al.* 2007, Zreda *et al.* 2008). Further, capture of thermal neutrons by ^{39}K produces ^{36}Cl via the ^{39}K(n,α)^{36}Cl reaction (Table 1.1; Fig. 1.7). The corresponding reaction cross-section (Chadwick *et al.* 2006) (Fig. 1.7) is about 0.1‰ of that of the ^{35}Cl(n,γ)^{36}Cl reaction; however, where potassium is a major element in target minerals its contribution will be significant.

At SLHL, stopped negative muons account for 6% and 11% of ^{36}Cl production from K and Ca, respectively (Stone *et al.* 1996c, Evans *et al.* 1997, Stone *et al.* 1998, Evans 2002), a value that increases rapidly with depth in the subsurface (Section 1.5).

There are considerable discrepancies in the published element-specific cosmogenic production rates, e.g. elemental production rate estimates of ^{36}Cl from Ca differ by a factor of two (Stone *et al.* 1996a, Swanson and Caffee 2001). Studies using whole-rock samples (Phillips *et al.* 2001, Swanson and Caffee 2001) tend to yield higher production rates than those using mineral separates (Stone *et al.* 1996a, Evans *et al.* 1997, Stone *et al.* 1998, Evans 2002). A recent study suggested that the higher production rates determined on whole-rock samples may be caused by underestimating the contribution of neutron-capture reactions of ^{35}Cl, which is often more abundant in the whole-rock samples (Schimmelpfennig *et al.* 2009). The total SLHL production rates for K, Ca and Fe, as determined on Cl-poor mineral separates, are about 171, 54 and 2 atoms ^{36}Cl g^{-1} a^{-1} (Stone *et al.* 1996a, Evans *et al.* 1997, Stone *et al.* 1998, Evans 2002, Stone 2005).

The thermal-neutron flux generated in rocks (Section 1.5) can also produce ^{36}Cl via the (n,γ) capture reaction (see above). This background concentration can be accounted for by calculating the rock's neutron flux, and the capture efficiency of ^{35}Cl in the sample, from the major and trace-element composition of the mineral used and/or the whole rock

(Phillips *et al.* 2001; Schimmelpfennig *et al.* 2009). The relative importance of this background diminishes with decreasing U, Th and Cl concentration in rocks, i.e. granites usually have a relatively high ^{36}Cl background, basalt and carbonates often have low (negligible) backgrounds.

Applications
^{36}Cl is the only cosmogenic nuclide that is routinely used for exposure dating of carbonate rocks. It is therefore regularly used in palaeoseismic studies, as fault scarps in carbonates are often well preserved (Zreda and Noller 1998, Mitchell *et al.* 2001, Benedetti *et al.* 2002, Palumbo *et al.* 2004; Section 4.1).

^{36}Cl is the only cosmogenic nuclide that can be used for basalts that do not contain He-retentive phenocryst phases, which could be used for ^{3}He-dating. Calibration efforts will continue to reduce the uncertainties in the element-specific production rates (e.g. Licciardi *et al.* 2008) that currently limit accuracy for the application of whole-rock samples (see above).

The combination of ^{10}Be in quartz and ^{36}Cl in feldspars from the same sample can be used to constrain complex exposure histories (Gosse and Phillips 2001).

Due to the high production rate of ^{36}Cl, the exposure dating of historic events is usually feasible when using Ca- and/or K- rich rocks/minerals, such as carbonates and K-feldspars.

2.3.5 Cosmogenic calcium (^{41}Ca)

Calcium has five stable nuclides (^{40}Ca, ^{42}Ca, ^{43}Ca, ^{44}Ca and ^{46}Ca); the cosmogenic radionuclide ^{41}Ca has a half-life of 104 ± 4 ka (Kutschera *et al.* 1992). It is currently not used for terrestrial exposure dating, but is routinely used to determine exposure histories and terrestrial ages of meteorites (Nishiizumi *et al.* 2000).

Meteoric/Atmospheric ^{41}Ca
Due to the lack of appropriate target elements in the atmosphere, production of atmospheric ^{41}Ca is negligible.

Production pathways and interferences
In silicate rocks, ^{41}Ca is mainly produced by thermal-neutron capture by ^{40}Ca, and to a lesser extent by spallation reactions from Ti, Cr and Fe (Nishiizumi *et al.* 2000, Audi *et al.* 2003).

The cross-section of the $^{40}Ca(n,\gamma)^{41}Ca$ reaction for thermal neutrons is about 1‰ of that of the corresponding $^{35}Cl(n,\gamma)\ ^{36}Cl$ reaction (0.41 and 44 barn, respectively; Chadwick *et al.* 2006). Considering their elemental concentrations and isotopic abundances of ^{40}Ca and ^{35}Cl (96.94 and 75.78%, respectively), and the different half-lives, the same procedures that are used to calculate ^{36}Cl production by thermal neutrons (Phillips *et al.* 2001) can be applied to ^{41}Ca. Using these procedures (Phillips *et al.* 2001) it can be calculated that granitic feldspar with 10% Ca would produce about 200 atoms $^{41}Ca\ g^{-1}\ a^{-1}$ (SLHL). The target nuclide (^{40}Ca) and the cosmogenic nuclide (^{41}Ca) cannot be separated chemically. Therefore, the maximum $^{41}Ca/Ca$ ratio that can be realized in a Ca-rich mineral is limited by the saturation concentration of ^{41}Ca in a material, and its Ca-concentration. In granites, the saturation $^{41}Ca/Ca$ ratio is $\sim 2 \times 10^{-14}$ (SLHL). Accurately measuring such low $^{41}Ca/^{40}Ca$ ratios is a challenge for most AMS installations (Section 2.4). At high altitudes, the ratio would increase with increasing cosmic-ray flux (Section 1.4). In carbonates, which have a higher cosmogenic thermal-neutron flux (Phillips *et al.* 2001), the maximum $^{41}Ca/Ca$ ratio would be about twice that of granite.

The analysis of ^{41}Ca produced by spallation reactions from Ti, Cr and Fe in Ca-poor minerals (such as Fe- and Ti-oxides) would not have the limitations discussed for production from Ca. This is because, for the spallogenic production pathways, the $^{41}Ca/Ca$ ratio can be increased by increasing the sample size (Section 2.3). Using the $^{41}Ca/^{36}Cl$ production ratio from Fe in meteorites, 1.10 ± 0.04 (Albrecht *et al.* 2000), and the ^{36}Cl production rate in iron (Stone 2005) as guidelines, it can be estimated that the ^{41}Ca production rate is around 2 atoms g^{-1} Fe a^{-1} (SLHL). Due to the smaller mass difference, the production rate from Ti should be significantly higher than that from Fe (Section 1.5).

Applications

Exposure dating of terrestrial iron–titanium oxides using ^{41}Ca should be feasible with current AMS technology. If used in conjunction with other cosmogenic isotopes (e.g. ^{10}Be in quartz from the same sample), spallogenic ^{41}Ca could constrain complex exposure histories.

2.3.5 Cosmogenic manganese (^{53}Mn)

Manganese has one stable nuclide (^{55}Mn); the cosmogenic radionuclide ^{53}Mn has a half-life of 3.7 ± 0.4 Ma (Honda and Imamura 1971). ^{53}Mn is a recent addition to the terrestrial cosmogenic toolbox (Schäfer *et al.* 2006, Gladkis *et al.* 2007).

Meteoric/Atmospheric ^{53}Mn

Due to the lack of appropriate target elements in the atmosphere, production of atmospheric ^{53}Mn is negligible (Schäfer *et al.* 2006).

Production pathways and interferences

^{53}Mn is produced by spallation reactions from Fe, Mn (Schäfer *et al.* 2006), Co and Ni. Production from Fe dominates by far, because of the small mass difference between Fe isotopes and ^{53}Mn, and because of the higher abundance of Fe relative to the other potential target elements, in most minerals.

At SLHL, stopped negative muons and fast muons together account for ~10% of ^{53}Mn production from Fe (Heisinger *et al.* 2002a), a value that increases rapidly with depth in the subsurface (Section 1.5). The SLHL production rate of ^{53}Mn in iron has been determined as 103 ± 11 atoms g^{-1} a^{-1} (Schäfer *et al.* 2006).

The main interference in the detection of ^{53}Mn by AMS is its stable isobar ^{53}Cr, and a considerable analytical effort is directed at eliminating and/or suppressing this interference (Lahiri *et al.* 2006, Schäfer *et al.* 2006, Gladkis *et al.* 2007). The ^{53}Cr interference is the reason why currently only large/powerful AMSs are able to analyse ^{53}Mn in terrestrial samples (Schäfer *et al.* 2006, Gladkis *et al.* 2007).

Applications

Due to its long half-life, ^{53}Mn can be used for exposure histories as old as 15 Ma (Schäfer *et al.* 2006), i.e. it can be used in most settings where stable cosmogenic nuclides previously had a unique advantage over the radionuclides. In principle, any iron-bearing material that is stable under Earth surface conditions can be used in conjunction with ^{53}Mn (Schäfer *et al.* 2006). For analytical reasons, Fe-rich minerals low in Mn and Cr will probably yield the best results.

2.4 Sample preparation

Sample preparation for cosmogenic nuclide analysis is aimed at: (i) concentrating and purifying the selected target material/mineral by physical and chemical preparation and (ii) chemically enriching the cosmogenic isotope and separating it from interfering isotopes of other elements.

2.4.1 Purification of mineral separates

Obtaining pure mineral separates is important for most cosmogenic nuclide applications. For instance, in the case of the stable cosmogenic nuclides (noble gases), impurities may not be able to retain noble gases; for example, a sample containing 5% non-retentive impurities may yield concentrations that are 5% too low (Bruno *et al.* 1997, Schäfer *et al.* 2006). In the case of radionuclides, impurities may contain interfering elements and/or meteoric components that disturb or invalidate the later measurements. In some cases, a few grains of an accessory mineral can compromise the reliability of results (e.g. beryl in the case of ^{10}Be; zircon or other U–Th rich minerals for ^{21}Ne). The only cases where mineral separation may not be necessary are for whole-rock samples that can be used as the target material (e.g. for ^{36}Cl). Despite the considerable effort that goes into mineral separation, it is usually a false economy to compromise on the purity of the mineral separates that are to enter the analytical process.

Physical sample preparation involves standard mineral preparation techniques such as crushing, sieving, density and/or magnetic separation, and sometimes handpicking of grains (the latter mostly for noble gas analysis). Physical–chemical separation techniques such as froth flotation can be used to separate mineral phases based on their wetting properties using organic surfactants (e.g. separating quartz from feldspar; Herber 1969, Vidyadhar *et al.* 2002). The combination of suitable techniques for a particular sample depends on the nature of the rock, which is a topic of relevant handbooks (Taggart 1945, Ney 1986) and is therefore not covered here.

Chemical sample purification using acids (selective etching) removes more easily soluble impurities, and can also remove meteoric components when a significant proportion of the target mineral is removed by the etching process (30% in the case of ^{10}Be in quartz, for example, Kohl and Nishiizumi, 1992). Density separation can be used to remove remaining acid-resistant impurities (e.g. zircon, garnet).

The clean mineral separates, stripped of mineral impurities and of adsorbed meteoric components, can then either enter the chemical procedure for AMS-target preparation (radionuclides) or are ready for noble-gas analysis (stable nuclides).

The amount of purified sample that is necessary for analysis depends on the nuclide analysed, the target mineral and the expected exposure age/erosion rate recorded by the sample. For ^{10}Be analysis, for example, *typical* sample weights range between 10 and 100 g of purified quartz (weight after physical preparation and chemical leaching). For noble gases, the sample sizes rarely exceed 1 g of pure mineral separate; commonly the samples are a few hundred mg, sometimes only tens of mg.

The weight of the sample collected in the field must reflect these needs, i.e. be sufficiently large to allow the preparation of an adequate amount of purified mineral separates.

2.4.2 Preparation of AMS targets

As with the mineral preparation (Section 2.4.1), there is no unique way to conduct the subsequent chemical preparation of targets for AMS measurements (Section 2.5). Many preparation labs have their distinct recipes that achieve the goal of delivering high-quality targets for subsequent AMS measurements. Examples of detailed procedures for ^{10}Be, ^{26}Al and ^{36}Cl published on the internet can be found at http://depts.washington. edu/cosmolab/ and http://www.uvm.edu/cosmolab/, provided by John Stone of the University of Washington and Paul Bierman at the University of Vermont, respectively. These sites also contain links to the procedures of other laboratories. Further representative references, by no means the complete list, are provided for specific nuclides below. While being different in detail, the various preparation recipes have important features in common that are outlined in the following.

Generally, the chemical procedures aim at a near-quantitative collection and isolation of the cosmogenic nuclide in question, the elimination of isobars (isobars are atoms and/or molecules of the same mass as the isotope/nuclide of interest) from the sample, and to finally render a material that is suitable for AMS measurements.

In cases where concentrations of the corresponding stable isotopes are too low to reliably form precipitates from sample solutions (i.e. the ^{9}Be, ^{27}Al and ^{35,37}Cl concentrations, in the case of ^{10}Be, ^{26}Al and ^{36}Cl, respectively), a carrier material is added just prior to, or immediately after, the dissolution stage. Complete mixing of sample and carrier material makes the subsequent procedure insensitive to incomplete recovery of the radionuclide because the radionuclide/stable nuclide ratio remains fixed. The purity of the carrier added, i.e. the native concentration of the relevant radionuclide (^{10}Be, ^{26}Al and ^{36}Cl), determines the minimum isotopic ratios that can be analysed with AMS (Section 2.5); ideally carriers should not contain any atoms of the corresponding radionuclide. In practice, carriers with radionuclide/stable nuclide ratios $<10^{-14}$ are suitable for most applications. The dating of young rocks, however, usually requires carrier isotopic ratios $<10^{-15}$ (Merchel *et al.* 2008a, Schaefer *et al.* 2009).

To give an idea about the labour-intensive preparation processes, essential features of the chemical preparation for the most commonly used cosmogenic radionuclides (^{10}Be, ^{26}Al and ^{36}Cl) are described below. Some nuclide-specific features for nuclides that are not discussed in this section (^{14}C, ^{41}Ca and ^{53}Mn) have already been discussed in the previous section (Section 2.3).

^{10}Be and ^{26}Al target preparation from quartz starts with the dissolution of the purified quartz in hydrofluoric acid. A ^{9}Be carrier is added prior to dissolution and, depending on the Al concentration of the quartz, an Al carrier may be added as well. The solution is evaporated to dryness, removing all silicon as gaseous SiF_4 in the process. The residue (fluorides of mostly Be, Al, Fe, Ti and alkali and alkali-earth elements) is subsequently taken up in hydrochloric acid. In several recipes this step is preceded by fuming of the original residue in an acid with a high-boiling point (perchloric or sulfuric acid), to destroy fluorides and drive off boron as BF_3 (^{10}B is an interfering isobar of ^{10}Be and needs to be removed during the target preparation). In subsequent steps, iron is removed as an anionic chloride complex on anion-exchange resin columns. Ti is removed either as precipitate (Ti hydroxide or oxide) or using cation-exchange resin columns. Cation-exchange resin columns are used to separate Al and Be from each other and from cations of the alkali metal (Li, Na, K, etc.) and alkali-earth metal groups (Mg, Ca, etc.).

Be is precipitated from the Be fraction as $Be(OH)_2$, rinsing of which removes remaining soluble $B(OH)_3$. $Be(OH)_2$ is then thermally decomposed (calcination) to BeO. Note that this step is often, wrongly, referred to as 'oxidation', however, no change in the oxidation state of Be, or any other compound, occurs. Finally, the BeO is mixed with metal (e.g. Ag or Nb) and pressed into cathodes for AMS measurements. An equivalent process is used to convert the Al fraction to AMS targets. For ^{26}Al, a separate determination of the stable Al content of the sample is required (Section 2.3.3). A selection of relevant references for ^{10}Be and ^{26}Al preparation techniques are Nishiizumi *et al.* (1989), Brown *et al.* (1991), Nishiizumi *et al.* (1991b), Kohl and Nishiizumi (1992), von Blanckenburg *et al.* (1996), Ochs and Ivy-Ochs (1997), Stone (1998), Merchel *et al.* (2008a) and aforementioned internet sites.

For ^{36}Cl target preparation, the main concerns are the removal of the isobar ^{36}S and preventing loss of Cl through (selective) volatilization in the preparation process. Samples are dissolved in acids (nitric acid for carbonates, a mixture of hydrofluoric and nitric acid for silicates), with particular care taken to avoid driving off Cl⁻ as HCl, which may occur during violent reactions (e.g. bubbling of CO_2 in the case of carbonates) and/or by using stronger acids than HCl. Cl carrier is added, either prior to or immediately after dissolution. Isotopically enriched Cl carrier (i.e. using stable chlorine with a ^{37}Cl/^{35}Cl ratio significantly different from natural Cl; Section 2.3.4) can be used to derive the Cl concentration of the sample from the AMS measurement, otherwise a separate determination of the stable Cl concentration is necessary. As soon as the sample solution and the carrier are homogenized, Cl loss is no longer a concern (as it would affect sample ^{36}Cl and carrier to the same extent, i.e. not change the ^{36}Cl/Cl ratio of the sample solution). Subsequently, chloride is precipitated as AgCl, and sulfate removed as $BaSO_4$. Dissolution of the AgCl precipitates with ammonia, repeated precipitation of sulfate as $BaSO_4$ and subsequent re-precipitation of chloride as AgCl, helps to remove sulfur as a significant source of interference. The purified AgCl is pressed into a target holder lined with AgBr for AMS analysis. Useful references on ^{36}Cl target preparation are, for example, Stone *et al.* (1994), Bierman *et al.* (1995), Stone *et al.* (1996a), Desilets *et al.* (2006a) and the aforementioned internet sites.

2.5 Analytical methods

Accelerator mass spectrometry (AMS) and static noble-gas mass spectrometry (NGMS) are routinely used to measure isotopic abundances of *in situ*-produced cosmogenic radionuclides and cosmogenic noble gases, respectively.

2.5.1 Noble-gas mass spectrometry

The principles of NGMS are similar to any conventional mass spectrometer (de Laeter 1998). Positive ions of the element of interest are produced in an ion source and accelerated to several kV (Nier 1947). The resultant ion beam is deflected in the field of a sector magnet, and ions are recorded as a function of their charge over mass ratio in Faraday cups and/or electron-multipliers (Nier 1947, Reynolds 1956, Clarke *et al.* 1969, de Laeter 1998).

Cosmogenic noble gases are released from samples by heating (cf. Niedermann 2002). Prior to analysis, the reactive gas species are removed by exposure to getter materials and the noble gases are separated from each other using cryogenic traps (Reynolds 1956, Clarke *et al.* 1969, Niedermann *et al.* 1993, de Laeter 1998).

NGMSs used for helium isotope analysis usually have a mass resolution that is sufficiently high to resolve the ^{3}He$^+$ from the T$^+$–HD$^+$ doublet (HD = molecule of a hydrogen and a deuterium atom; T = tritium), and have a high abundance sensitivity that allows measurements of ^{3}He/^{4}He ratios as low as $\sim$10^{-8} (Clarke *et al.* 1969). These are amongst the smallest isotope ratios measured with conventional static gas mass spectrometers.

The doubly charged ^{40}Ar^{2+} and CO$_2{}^{2+}$, which interfere with the measurement of ^{20}Ne$^+$ and ^{22}Ne$^+$, respectively, are suppressed using suitable getters and cold traps. These interferences can be further reduced by partial resolution of ^{40}Ar^{2+} from ^{20}Ne$^+$ (Niedermann *et al.* 1993) or by lowering the ionization potential in the ion source (Bruno *et al.* 1997).

2.5.2 Accelerator mass spectrometry

The high acceleration voltages available in AMS, typically between 1–10 MV (Hellborg and Skog 2008), permit an extremely good discrimination against isobaric, isotopic and molecular interferences (Hellborg and Skog 2008). Isotopic ratios as low as 10^{-16} can be measured; 10^{-15} is commonly achieved (Kutschera 2005).

A *typical* AMS system includes the following (after Hellborg and Skog 2008):

(i) Production of negative ions from a target material (i.e. the sample) by Cs-sputtering in an ion source. Several important interfering isobars do not form negative ions. For example ^{14}N and ^{26}Mg are effectively suppressed during analysis of ^{14}C and ^{26}Al, respectively.

(ii) Acceleration of negative ions from ground potential to a high positive voltage.

(iii) Recharge of all ions to positive by stripping off electrons during the passage through a stripper foil or gas in the centre of the accelerator. Positive molecular ions, which may include potential isobaric interferences, are dissociated. At a charge state $>2+$, molecular ions are essentially absent.

(iv) Acceleration of the now-positive ions back to ground potential.

(v) Removal of unwanted ions (such as the fragments of the dissociated molecules) by deflection of the ion beam in electric and magnetic fields.

(vi) Final separation, identification and counting of individual isotopes are based on their residual energy.

Most current AMS facilities have acceleration voltages between 3 and 5 MV (Kutschera 2005). High energies facilitate separation of ^{36}Cl from interfering ^{36}S (Kutschera 2005). Current developments in AMS technology are bringing low-energy AMS (<1 MV) within reach for terrestrial cosmogenic *in situ* applications (^{10}Be, ^{26}Al), but those applications are still experimental (Suter 2004, Kutschera 2005). For radiocarbon dating, such low-energy AMSs are already routinely used (Suter 2004, Kutschera 2005, Hellborg and Skog 2008).

3

Production rates and scaling factors

The accurate knowledge of production rates of cosmogenic nuclides, and how they change as a function of latitude and altitude, is arguably the most important requirement for their successful application to Earth surface sciences. Exposure ages or erosion rates calculated from cosmogenic nuclide concentrations (Chapter 4) can only be as accurate as the production rates and scaling factors they rely on. The aim of this chapter is therefore to provide some background to understand underlying methods and to illuminate key aspects of the ongoing scientific discussion on scaling factors.

3.1 Deriving production rates

Production rates can be derived via three different methods from: (i) physical principles, (ii) irradiation experiments and (iii) geological calibration.

From physical principles

Using cosmic-ray flux and nuclear reaction cross-section data the cosmic ray cascade in the atmosphere and its interaction with materials on the Earth's surface can, in principle, be modelled by numerical simulations (Masarik and Reedy 1994, 1995, 1996, Masarik and Beer 1999). In practice, the accuracy of this approach is hampered by the limited knowledge of the nuclear reaction cross-sections involved, particularly for neutron-induced reactions (Masarik and Reedy 1995). For proton-induced reactions, a considerable body of data is available (Masarik and Reedy 1995, Michel *et al.* 1996, Leya *et al.* 1998); for neutrons, it is usually assumed that neutron-induced reactions have cross-sections similar to those of proton-induced reactions (Masarik and Reedy 1995). However, these approximations have uncertainties of the order of 25% (Masarik and Beer 1999, Leya *et al.* 2000, Sisterson 2005). The large uncertainties in the fundamental parameters used in the transport codes currently limit direct application of these models to *in situ* applications, such as for production-rate determinations.

From irradiation experiments

In irradiation experiments, pure elements or chemical compounds are either exposed to artificial muon, proton or neutron beams (Leya *et al.* 1998, Heisinger *et al.* 2002a, Heisinger *et al.* 2002b, Sisterson 2005, Nishiizumi *et al.* 2009), or to environmental cosmic rays (Yokoyama *et al.* 1977, Nishiizumi *et al.* 1996, Brown *et al.* 2000, Vermeesch *et al.* 2008).

Beam-line experiments are the key source of the reaction cross-section data required for the numerical calculations discussed above. At the time of writing, promising new data is emerging from rare neutron-beam experiments (Nishiizumi *et al.* 2009), which in due course should improve the accuracy of models. Muon-beam experiments currently provide the most widely used values on the contribution of muons to the production of cosmogenic nuclides (Heisinger *et al.* 2002a, Heisinger *et al.* 2002b) (Section 1.5).

Experiments that rely on the environmental cosmic-ray flux either aim to describe relative changes of the production rates as a function of altitude and latitude (Brown *et al.* 2000, Vermeesch *et al.* 2008) (Section 3.2), or at constraining nuclide production rates for use with geological samples (Nishiizumi *et al.* 1996). Due to the relatively short time-scale of these experiments, compared to the 11-year solar cycle that modulates the cosmic-ray flux (Section 1.2), estimates of long-term average solar activity have to be used to derive long-term average production rates that are applicable to geological samples (Nishiizumi *et al.* 1996).

From geological calibration

The most commonly applied method for deriving cosmogenic nuclide production rates is to measure cosmogenic nuclide concentrations in samples from stable, continuously exposed, geological surfaces. Either surfaces of known age, or surfaces sufficiently old that radionuclides can reach saturation, are used (Gosse and Phillips 2001). The overwhelming majority of production-rate estimates for spallogenic production that were discussed in the previous chapter were derived by geological calibration. Ideally the nature of the surface used for geological calibration should be such that significant erosion, prior exposure and/or intermittent burial can be excluded with confidence (Chapter 4).

Depending on the nature of a surface (glacial feature, lava flow etc.; see also Chapter 4), independent age constraints may be derived from diverse techniques such as radiocarbon-, luminescence- and Ar/Ar-dating, as well as dendrochronology and varve chronology (e.g. Kurz *et al.* 1990, Cerling and Craig 1994a, Kubik *et al.* 1998, Dunai and Wijbrans 2000, Balco and Schaefer 2006). At its simplest, a cosmogenic nuclide production rate is calculated from the measured nuclide concentrations in a sample, divided by the surface's independently determined age. For radionuclides, the concentrations also have to be corrected for decay. It should be evident that the production rate from a surface cannot be more accurate than the independent age constraint(s) used. Note, however, that this principal limitation is regularly overlooked in practice, when multiple production-rate determinations are averaged. Specifically, this occurs when means from multiple production-rate determinations, derived from a single site with a single independent age constraint, are calculated. If a sufficiently large number of production rates are averaged in this way, the error of the mean production rate may become apparently smaller than the uncertainty of the independent age constraint; this is logically wrong. The correct procedure is to calculate the mean nuclide concentration from multiple samples (this value will become more precise with increasing number of observations; i.e. concentration determinations) and divide the mean concentration by the independently derived age.

When samples reach saturation, the rate of radioactive decay is identical to the cosmogenic production rate (Section 2.1). In this case, the production rate of the nuclide is the product of the nuclide's saturation concentration and its decay constant. Therefore, if a sufficiently high age of a surface can be established independently (e.g. using the aforementioned dating techniques, or using longer-lived cosmogenic nuclides), the production rate is straightforward to calculate (Nishiizumi *et al.* 1989, Jull *et al.* 1992, Brook *et al.* 1995b, Lifton 2008).

The local production rates, as obtained by geological calibration, are usually also reported normalized to sea-level and high latitude (SLHL). Consequently, these normalized values are a function of the scaling factors used (Section 3.2). The normalization of SLHL production needs to be consistent with the scaling factors used for an application (Balco *et al.* 2008).

3.2 Scaling factors

Scaling factors describe the variability of the cosmic-ray flux, as relevant to cosmogenic nuclide production, as a function of altitude and latitude. In other words, they are a formulaic description of the influence of the Earth's and Sun's magnetic fields, and atmospheric mass on cosmogenic nuclide production rates (Sections 1.2 and 1.3). Several scaling factors also describe the effects for the secular variation of these parameters (Section 3.3.2). Scaling factors are the essential tools needed to translate the local production rates derived from geological calibration sites to any location where cosmogenic nuclides are applied to address geological questions. Normally, scaling factors are formulated such that the product of the site-specific scaling factor and the SLHL production rate of a nuclide provides the site-specific nuclide production rate.

Similarly to production rates, scaling factors may be derived from numerical simulations based on physical principles (Masarik and Beer 1999, Masarik *et al.* 2001), or from geological calibrations (Lifton 2008). Further, proxy measurements, in lieu of geologic calibrations, are commonly used to derive scaling factors (Lal 1991, Dunai 2000, Stone 2000, Dunai 2001a, Desilets and Zreda 2003, Lifton *et al.* 2005, Lifton *et al.* 2008). A combination of these methods is, of course, possible, and is valuable for cross-calibration and assessing consistency between methods.

Numerical simulations (Masarik and Beer 1999, Masarik *et al.* 2001) agree with neutron-monitor survey measurements to within 10% (Masarik and Beer 1999). They appear to be less sensitive to the geomagnetic field, since they show a less pronounced latitude effect (Section 1.2) than neutron monitors (i.e. 10% smaller). This may be due to the uncertainty of the underlying parameters (Section 3.1), and/or an inherently different energy bias of the two approaches (Section 1.6). In any case, this indicates that, currently, results derived by the two methods should not be combined in practical use (e.g. using a numerically derived geomagnetic correction in conjunction with a neutron-monitor-based scaling factor; see below). Generally, as maintained throughout this chapter, it is prudent to keep consistency, and not to combine approaches that superficially may appear similar. Do not combine approaches that do not give the same answer to the same question (e.g., how large is the latitude effect?).

Hitherto it has been difficult to assemble a global network of surfaces suitable for geological calibration that could be used as a grid to construct scaling factors (Section 3.3). An ongoing effort to use surfaces saturated with ^{14}C (Lifton 2008) may change this in the near future. Logically,

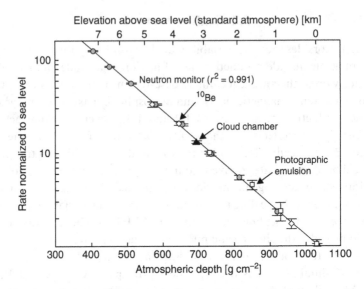

Fig. 3.1. The normalized counting/production rates of neutron monitor (grey circles), photographic emulsion (white squares), cloud chamber (black triangles) and ^{10}Be water target experiments (white squares) are shown as a function of atmospheric depth (which is a function of elevation). The normalization is relative to sea level. All four types of proxy data behave, within their uncertainties, the same to changes in cosmic-ray flux. Observational data collected at locations with cut-off rigidities between 0.5 and 4 GeV; and during periods with a solar modulation between 500 and 700 MeV. Format from Dunai (2001), data from Teucher (1952), Brown (1954), Sandström (1958), and Brown *et al.* (2000).

geological calibrations are the ultimate test for any scaling factor, since they relate perfectly to one of the technique's prominent applications: the dating of geological surfaces. Where available in significant numbers, geological calibrations have already been used to assess scaling factors (Dunai 2001a, Balco *et al.* 2008).

Proxy measurements for cosmogenic nuclide production rates are mostly conducted with neutron monitors and photographic emulsions (Section 1.6). All currently used scaling factors for *in situ* production rely partially (Lal 1991, Stone 2000) or entirely (Dunai 2000, 2001a, Desilets and Zreda 2003, Lifton *et al.* 2005) on neutron monitor data. The scaling factors of Lal (1991) also rely on photographic emulsion data. Diverse proxy measurements conducted with photographic emulsions, neutron monitors, ^{10}Be production in water targets and cloud chambers *can* provide consistent results (Fig. 3.1), which may be taken as a sign of their robustness against experimental and instrumental bias. Care has to be taken that the response of proxy measurements to the cosmic-ray energy spectrum (Section 1.1) is proportional to the cosmogenic nuclides of interest (Dunai 2001b). If this is the case, the various observations can be normalized to a specific location, e.g. SLHL, and combined for use in scaling factors. At the time of writing, it appears that proxy-derived scaling factors describe the spallogenic production of cosmogenic nuclides (Section 1.5) adequately (Balco *et al.* 2008). With further improvements in accuracy of geological calibrations, e.g. from results of the CRONUS networks, a more subtle assessment should become possible.

3.3 Building scaling factors

There is relatively little principal difference between the data that lie at the basis of currently used scaling factors (Lal 1991, Dunai 2000, Stone 2000, Dunai 2001a, Desilets and Zreda 2003, Lifton *et al.* 2005). The purely neutron-monitor-derived scaling factors (Dunai 2000, 2001a, Desilets and Zreda 2003, Lifton *et al.* 2005) may differ in the amount of data used, but not in the type of data. Thus it is not entirely surprising that fundamental values, such as the amplitude of the latitude effect and attenuation coefficients at a given latitude, agree within stated uncertainties, when calculated in a consistent manner (Dunai 2000, 2001a, b, Desilets and Zreda 2003).

Also, the scaling factors of Lal (1991) and Stone (2000) use, for instance, the same neutron monitor data as Dunai (2000, 2001) at sea level. However, the former two rely on a combination of different neutron-monitor configurations, which have different energy responses, to assess the altitudinal variation of the cosmic-ray flux (Lal 1958; Lifton personal communication). The latter gives rise to the use of different attenuation coefficients, as compared to the other scaling factors (Dunai 2000, 2001a, Desilets and Zreda 2003, Lifton *et al.* 2005), which is the only significant data-driven difference between the various methods of scaling.

Aside from the different attenuation coefficients used by Lal (1991) and Stone (2000), the differences between scaling factors are mostly driven by the manner in which the grids that arrange observational data are constructed. It largely boils down to the choice of how atmospheric pressure and the geomagnetic field are described, and whether, and which, secular variations are taken into account.

In the following, the design features shared by the various scaling factors, and those that set them apart, are briefly introduced. This introduction focuses on the nucleonic component, which causes spallation reactions, as this component dominates the production of cosmogenic nuclides (Chapter 2). For a comprehensive discussion, I refer the interested reader to the corresponding publications (Dunai 2001a, b, Desilets and Zreda 2003, Lifton *et al.* 2005, Lifton *et al.* 2008). At the time of writing, the scientific discussion on scaling factors, and their design, is still ongoing.

3.3.1 Coordinate systems

A key task for scaling factors is to translate the geographic coordinates of a sampling site, such as altitude and latitude, into coordinates relevant to cosmic-ray phyics and cosmogenic-nuclide production. Such a translation should ideally be as simple as possible, to remain practical, and as complicated as necessary, to accurately describe the relative changes in nuclide production rates. A suitable coordinate system allows the arrangement of the normalized proxy measurements in a grid-like form, and the accurate calculation of interpolated values between them, thereby generating the mathematical formulations that are the scaling factors. It is implicitly assumed that the same parameters affecting the present-day proxy measurements have equally affected the time-integrated nuclide production in geological samples.

Geographic coordinates

The scaling factors of Lal (1991) use geographic coordinates, altitude and latitude, to describe variations in the cosmic-ray flux. Two assumptions are made to achieve this: (i) the magnetic field at the time of collection of the proxy measurements is a geocentric axial dipole field, i.e. that geomagnetic and geographic latitude are identical and (ii) the atmospheric pressure varies with altitude according to a standard relationship that is applicable globally. Implicitly, these assumptions prescribe that the same conditions apply to the exposure history of geological samples. If these conditions are valid, geomagnetic and geographic coordinates are equivalent/interchangeable, as is commonly assumed in the practical application of Lal's (1991) scaling factors.

Stone's (2000) scaling factors are a reformulation of Lal's (1991) scaling factors, which express altitude as atmospheric pressure, i.e. eliminate the assumption of a standard altitude–pressure relationship, and replace them with actual values. The first assumption implicitly remains in place.

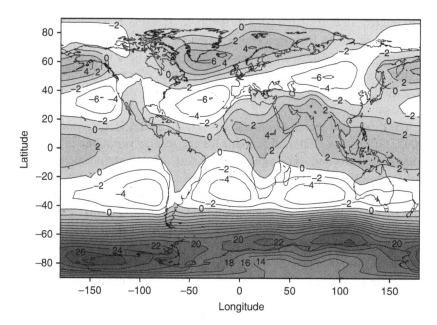

Fig. 3.2. Differences (in %) in sea-level cosmogenic nuclide production rates calculated for actual sea-level pressure relative to those that would be expected when assuming a uniform sea level pressure of 1013.25 mbar (standard atmosphere). Positive values occur in regions with low mean pressure, negative in those with high pressure. The contours are calculated for the average global pressure field averaged over the last 50 years (Allan and Ansell 2006), using the scaling factors of Dunai (2000, 2001). For the calculation geographic coordinates were used rather than geomagnetic field models. Therefore the figure should not be used for actual production rate corrections; it should serve to illustrate the qualitative effect of long-term pressure anomalies. Format after Stone (2000).

All scaling factors succeeding Lal (1991) use pressure as the coordinate for the vertical (Fig. 3.2).

Geomagnetic coordinates

The scaling factors of Dunai (2000) use the actual geomagnetic latitude (expressed as magnetic-field inclination) and the actual atmospheric pressure to describe the variation in cosmic-ray flux. Magnetic-field strength variations are not incorporated in these scaling factors. Thus, similarly to the scaling factors of Lal (1991) and Stone (2000), it is only valid for the present-day magnetic-field strength.

The revised scaling factors of Dunai (2001a) explicitly include variations in magnetic-field intensity in the description of the cosmic-ray flux.

The analytically calculated cut-off rigidity R_C (Section 1.2) and the actual atmospheric depth (i.e. pressure) are used as parameters to describe the cosmic-ray flux. The scaling factors of Desilets and Zreda (2003) and Lifton *et al.* (2005) use trajectory-tracing-derived effective cut-off rigidity R_{CE} (Section 1.2) and actual atmospheric depth as parameters to describe the cosmic-ray flux.

The three most recent scaling factors (Dunai 2001a, Desilets and Zreda 2003, Lifton *et al.* 2005) all account for the secular variations in atmospheric pressure and geomagnetic field that affect cosmogenic nuclide production (Section 3.3.2; Fig. 3.3).

The scaling factors of Desilets and Zreda (2003) and Lifton *et al.* (2005) also incorporate effects of solar modulation (Section 1.2) that modulate the primary cosmic-ray flux. The scaling factors of Dunai (2000, 2001a) are representative for the long-term average solar modulation of 550–620 MeV (Michel *et al.* 1996, Masarik and Beer 1999, Dunai 2001b). The mean Holocene solar modulation, ~420 MeV (Steinhilber *et al.* 2008), appears to be lower than the long-term modulation (see also the section on solar modulation at the end of this chapter).

Another difference in the coordinates used (Dunai 2001a, Desilets and Zreda 2003, Lifton *et al.* 2005) is how the cut-off rigidities are derived, and how cut-off rigidities are calculated for the geological past. For instance, at sea-level, trajectory-tracing-derived R_{CE} allow a very accurate fitting/prediction of modern observations with errors of ±0.3% (Section 1.2); whereas the corresponding errors of analytically calculated R_C are ±2% in dipole-dominated fields (>70% dipole; see Section 1.2). Fitting modern observations, R_{CE} clearly delivers the best results; the larger uncertainty for R_C arises from the fact that penumbral effects are not accounted for (Section 1.2). For the same reason, numeric values for R_{CE} and R_C are usually different for the same location (Section 1.2). It is, however, important to note that accurate high-order descriptions of the geomagnetic field are required to perform trajectory tracing calculations accurately (Bhattacharyya and Mitra 1997, Lifton *et al.* 2008). Consequently, the uncertainties involved in calculating R_{CE} will increase when going into the geological past, for which such accurate geomagnetic models do not exist at present (Knudsen *et al.* 2008, Valet *et al.* 2008). At some point, all scaling factors (have to) assume a geocentric axial dipole field (i.e. prior to 10–20 ka).

To bridge the temporal and methodological gap between modern trajectory-derived values and trajectory-derived values appropriate for long-term mean geomagnetic conditions (i.e. dipole), Desilets and Zreda (2003)

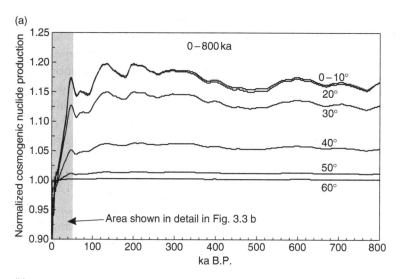

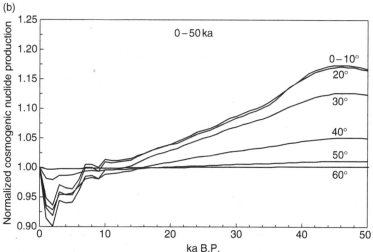

Fig. 3.3. Relative changes in cosmogenic nuclide production rates at sea level in response to changes of the Earth's magnetic field strength (Sint-800 record of Guyodo and Valet (1999). Figure a covers the time between the present and 800 ka, figure b shows the changes in the last 50 ka. These changes were calculated for the scaling factors of Dunai (2001); however, production rates derived with alternative time-variable scaling factors (Desilets and Zreda 2003, Lifton *et al.* 2005) would behave in a similar fashion (Balco *et al.* 2008). Figure from Dunai (2001).

perform trajectory tracing in synthetic dipole fields of various intensities. This approach implicitly assumes that the details of the penumbral effects of the synthetic dipole field are equivalent to the time-averaged penumbral effects of the actual geomagnetic field. The associated uncertainty is not provided by the authors. Lifton *et al.* (2005) take a different approach using a best-fit function of a latitude vs. R_{CE} distribution, calculated for a 1955 field model, to approximate R_{CE} in a dipole field. This procedure introduces a maximum uncertainty of $\pm 6\%$ to production rates at sea level and high R_{CE} (Lifton *et al.* 2005).

The expanded dipolar analytical equation for calculating R_C (Eqn. (1.2)), used by Dunai (2001a), does not account for penumbral effects (Section 1.2) in the present geomagnetic field, and performs the same with any averaged dipole field of the geological past. The 'penalty' for this simplification (i.e. $\pm 2\%$, see above) occurs once and thus does not increase further when applied consistently to the geological past (i.e. no mixed use of parameters R_{CE} and R_C; Section 1.2).

3.3.2 Input parameters

The scaling factors that allow the use of actual values for atmospheric pressure (Dunai 2000, Stone 2000, Dunai 2001a, Desilets and Zreda 2003, Lifton *et al.* 2005), and that describe changes in the geomagnetic field (Dunai 2001a, Desilets and Zreda 2003, Lifton *et al.* 2005) and the solar modulation (Desilets and Zreda 2003, Lifton *et al.* 2005), give considerable freedom and responsibility to the user. The freedom pertains to the ability to describe the actual situation, and the responsibility to use the most appropriate values and/or models in a consistent and transparent manner. In this section I will discuss some of the choices that can be made, their uncertainties, and the potential influence these uncertainties may have on the accuracy of production rates. The users of scaling factors should always reveal *all* input parameters and data sources that were used to derive production rates in publications.

Atmospheric pressure

One of the largest and probably most persistent pressure anomalies on Earth is located above Antarctica, where pressures at all elevations are 20–40 mbar lower than standard atmosphere values (Stone 2000). The pressure drop with altitude over the Antarctic surface is influenced by airflow across the ice sheet and a persistent temperature-inversion layer (Stone 2000). Given the inherent link of the pressure anomaly to the presence of the ice-sheet it is likely that it has been a persistent feature since the onset of continental-scale ice-sheets in Antarctica. Ignoring this pressure anomaly, i.e. applying the standard atmospheric pressure–altitude relationship in Antarctica, introduces a systematic error of 20–30% in calculated ages, as production rates are 20–30% higher in Antarctica than at equivalent altitudes on the Northern Hemisphere (Stone 2000) (Fig. 3.2). Stone's (2000) suggestion of using an appropriate pressure–altitude relationship for Antarctica is now universally applied to pertinent cosmogenic studies there.

Long-term pressure anomalies are not limited to Antarctica (Dunai 2000, Stone 2000) (Fig. 3.2), and, at least for Holocene exposure histories, it is prudent to incorporate any long-term pressure anomalies into the production-rate calculations (Licciardi *et al.* 2008).

Over longer timescales, i.e. those covering glacial periods, the magnitude and position of pressure anomalies change with climate (Ackert *et al.* 2003, Staiger *et al.* 2007). Mechanisms involved include: (i) changes in global atmospheric dynamics, because ice sheets displace atmospheric mass (Staiger *et al.* 2007) and shifting climate zones (Ackert *et al.* 2003); (ii) quasi-stationary zones of low surface pressure at ice-sheet margins due to katabatic winds (Stone 2000) and (iii) atmospheric compression due to cooling (Dunai 2000). In areas outside Antarctica, these mechanisms can change local instantaneous production rates by up to 10% between glacial maximum and present-day conditions (Staiger *et al.* 2007). Long-term average effects are obviously smaller (Staiger *et al.* 2007).

The proposed mechanisms for pressure-anomaly corrections are all sound and physics based; however, their quantitative expression is dependent on the accuracy of the climate and/or ice-sheet models used to reconstruct past situations. It is, therefore, sensible to estimate the uncertainty of models (e.g. by using the range of results obtained by end-member models) to derive the uncertainty of production rates. For example, if there is a 30% uncertainty on a 10% effect, one would expect that the resulting production rates have an additional uncertainty of 3%. Depending on the overall uncertainties involved, and on the desired accuracy of the argument(s) based on the data, the pressure correction would probably be worthwhile in this example, despite the considerable model-uncertainty assumed.

Geomagnetic field

The effect of changing the time-integrated geomagnetic field strength is equivalent to changing the time-integrated geomagnetic latitude of a sampling site (Nishiizumi *et al.* 1989, Clark *et al.* 1995). This finding has been used to account for magnetic-field changes in conjunction with scaling factors that nominally do not allow for geomagnetic variations (Clark *et al.* 1995, Gosse and Phillips 2001, Balco *et al.* 2008). Depending on the geomagnetic latitude and altitude, magnetic-field strength variations can affect the instantaneous production rates by 35–60%, for low-latitude sites at sea level and high altitude, respectively (Dunai 2001a) (Fig. 3.4). Similarly, changes in the position of the geomagnetic pole (Ohno and Hamano 1993, Merrill *et al.* 1998) are equivalent to changes in the geomagnetic latitude (Licciardi *et al.* 1999, Dunai 2000, 2001a, Gosse and Phillips 2001). Depending on the latitude and altitude of a site, changes in instantaneous production rates can be as high as 15–35% (Dunai 2001a) (Fig. 3.4). Naturally the time-integrated effects of field-strength and pole-position variability will be smaller than the instantaneous effects. After *c.* 10–20 ka the time-integrated pole position is invariant and geocentric axial (Merrill *et al.* 1998). The time-integrated dipole strength, however, changes significantly (i.e. causing >3% changes in time-integrated production rates) until at least ~100 ka (Guyodo and Valet 1999, Dunai 2001a) (Fig. 3.3).

These geomagnetic-field reconstructions may have considerable uncertainties. Estimates for the average virtual axial dipole moment (VADM, see e.g. Merrill *et al.* 1998), for time intervals prior to the Holocene, commonly have uncertainties in the order of 10–50%, depending on the time interval and methods used (Guyodo and Valet 1999, Knudsen *et al.* 2008, Valet *et al.* 2008). Time-integrated values have smaller uncertainties. High-resolution Holocene records may have smaller uncertainties and often overlap/agree with historic records (Korte and Constable 2005, Knudsen *et al.* 2008). However, there is no general agreement on the nature of the Holocene magnetic field in general, and on its high-order reconstructions (Korte and Constable 2005, Knudsen *et al.* 2008, Valet *et al.* 2008).

Different geomagnetic records often disagree on details on the time-scale of several thousand years; however, they agree reasonably well on the overall trend e.g. that the dipole moment has increased gradually over the past 40 ka (Knudsen *et al.* 2008). Thus, they probably reliably reconstruct long-term changes in the magnetic field (Guyodo and Valet 1999). Differences between the records are related to differences in temporal

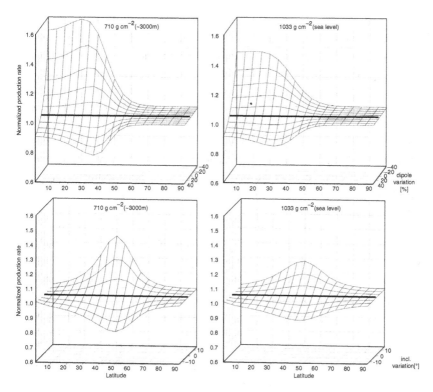

Fig. 3.4. Relative sensitivity of locations at different latitudes and altitudes (sea level and 3000 m elevation shown) to changes in cosmogenic nuclide production rates arising from changes in dipole strength and palaeolatitude. (Paleo-)latitude (λ) is related to inclination (I) by the relationship $\tan(I) = 2 \tan(\lambda)$; the bottom diagrams show sensitivity to changes in local field inclination, as are inclination values that are usually reported in palaeomagnetic studies. The figures were calculated using the scaling factors of Dunai (2001); however, production rates derived with alternative time-variable scaling factors (Desilets and Zreda 2003, Lifton *et al.* 2005) would behave in a qualitatively similar fashion. Figure from Dunai (2001).

resolution and to the nature and spatial distribution of the records (Knudsen *et al.* 2008, Valet *et al.* 2008). Also important are the normalization procedures used. The normalization of the 800 ka SINT-800 record (Guyodo and Valet 1999) was geared toward reproducing 5 ka-wide calibration intervals over the last 40 ka, whereas the normalization of the 2 Ma SINT-2000 record was geared to match averages of 100 ka calibration intervals (Valet *et al.* 2005), resulting in significantly different means (Valet *et al.* 2005). Currently, the SINT-800 is widely used for production-rate calculations (Balco *et al.* 2008).

The above discussion should have made it clear that the reconstruction of past geomagnetic field parameters is an active field of research; there is no model/record that has gained general acceptance and that could thus be prescribed. The abovementioned records are only a small subset of those published. Several records/models have already been used repeatedly for cosmogenic production-rate calculations (Ohno and Hamano 1993, Guyodo and Valet 1999, Yang *et al*. 2000), or are implemented in on-line calculators (Guyodo and Valet 1999, Korte and Constable 2005, Balco *et al*. 2008). This prior use does not necessarily signify precedence of these records over others; however, it is worth considering comparing results that may be obtained by the use of other records with one of the more frequently used ones (e.g. in the form of a robustness test). If there is no clear indication of why alternative geomagnetic records would yield more reliable results, comparability to results of other studies may be lost when exclusively using 'exotic' records, without necessarily gaining accuracy. For comparisons of results from different scaling factors, the same geomagnetic records/models should be applied consistently in order to make such an exercise meaningful (Balco *et al*. 2008). Generally, changing the geomagnetic (or atmospheric) model from that used in the original (calibration) studies requires the recalibration of the SLHL production rates from the original datasets using the same assumptions. This principle is implemented in the CRONUS web calculator (Balco *et al*. 2008).

To estimate the uncertainties associated with the time-integrated uncertainty of geomagnetic-field reconstructions, diagrams such as those provided in Fig. 3.4 can be used. These allow: (i) the delineation of regions where field changes and/or uncertainties matter, as well as (ii) the quantification of the changes in production rates as a function of relative changes/uncertainties in geomagnetic-field parameters.

Solar modulation

Solar modulation (Section 1.2) is highly correlated to solar activity, which in turn is related to the observable sunspot number (Masarik and Beer 1999, Wiedenbeck *et al*. 2005). The dominant 11-year sunspot cycle is modulated by the 22-year Hale cycle (Mursula *et al*. 2002), the 88-year Gleissberg, the 211.5-year Suess and the 2115-year Hallstatt cycles (Damon and Jirikowic 1992). This has caused fluctuations in the mean solar activity over the Holocene. Several periods during which solar

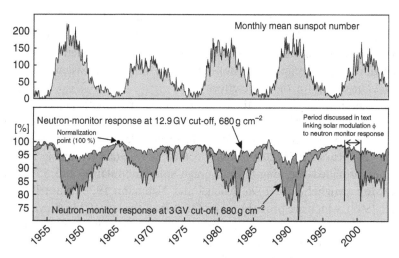

Fig. 3.5. Monthly mean sunspot number and neutron monitor response for the period between 1953 and 2004. The sunspot number is a qualitative measure for the solar activity; the relationship is not strictly linear (Wiedenbeck *et al.* 2005, Muscheler *et al.* 2007). The neutron monitor response at different cut-off rigidities is very similar in pattern, though not in amplitude. Sites with low cut-off rigidity (i.e. high latitude) show a stronger response to solar modulation. The neutron monitor response is normalized to the monthly mean intensity of May 1965. Sunspot and neutron monitor data, and their normalization, are from Lifton *et al.* (2005). The period used to link the actual solar modulation parameter Φ to neutron monitor response is from Wiedenbeck *et al.* (2005). For discussion see text.

activity was considerably reduced relative to present day have been identified during the Holocene e.g., the Maunder, Spoerer, Wolf, Oort and Dalton minima (cf. Lifton *et al.* 2005).

The actual value of the *average* solar modulation over the Holocene is currently contended within the framework of climate-change discussions. Some see the present-day activity as unusually high (Solanki *et al.* 2004); others see it as slightly elevated, but not unusual (Muscheler *et al.* 2007). As with geomagnetic reconstructions, the differences arise from the different data and normalization procedures used (Solanki *et al.* 2004, Muscheler *et al.* 2005, Solanki *et al.* 2005, Muscheler *et al.* 2007). The issue of the average Holocene solar modulation, and which records describe it best, may therefore be described as currently undecided. Reconstructions of past solar modulations are crucially dependent on magnetic-field-strength reconstructions (Solanki *et al.* 2004, Muscheler *et al.* 2005,

Solanki *et al.* 2005, Muscheler *et al.* 2007). Thus if a scaling scheme is used that allows the consideration of past fluctuations, the same magnetic reconstructions should be used as were used for the solar activity reconstructions (or a compatible one). This is to remain consistent, and to avoid correcting for the same effect twice.

During the past 50 years, the solar modulation parameter ϕ has ranged between 300 and 1200 MeV, depending on the solar activity (Masarik and Beer 1999, Wiedenbeck *et al.* 2005). The mean solar modulation over the last ~50 years (1953–2000) was ~650 MeV (Steinhilber *et al.* 2008). Values provided for long-term average solar modulation are 550 and 620 MeV (Michel *et al.* 1996, Michel and Neumann 1998, Masarik and Beer 1999), which are probably valid for at least the last 10 Ma (Leya *et al.* 2000). Thus it appears that the present-day mean solar modulation is rather similar to the long-term mean.

Due to the short periodicity of the main carrier frequency (1/11a) large magnitude fluctuations average out quickly. The longer-term modulations of the Gleissberg, Suess and Hallstatt cycles may indicate that solar modulation requires several hundred, or even several thousand years to attain the long-term average.

The effect of solar modulation on instantaneous cosmogenic-nuclide production between solar activity maxima and minima is in the order of ~18% at high latitudes and ~4% at low latitudes (Section 3.1; Fig. 3.5). These values apply to periods in which the solar-modulation parameter changed between ~300 and ~1100 MeV (Wiedenbeck *et al.* 2005). The response of the terrestrial neutron flux is approximately linear to the solar-modulation parameter (Wiedenbeck *et al.* 2005). Thus, for solar-modulation parameters within e.g. ~25% of the modern mean (i.e. 650 ± 160 MeV; mean of Steinhilber *et al.* 2008, see above) cosmogenic neutron fluxes will vary about $\pm$ ~3% at high latitudes, and $\pm$ ~0.8% at low latitudes.

4

Application of cosmogenic nuclides to Earth surface sciences

As introduced in Chapter 2, the application of cosmogenic nuclides to Earth surface sciences is a relatively recent affair. In 1986/87 terrestrial *in situ* cosmogenic nuclides came of age, with eight seminal papers reporting cosmogenic ^{3}He, ^{21}Ne, ^{22}Ne, ^{10}Be, ^{26}Al and ^{36}Cl in terrestrial rocks (Craig and Poreda 1986, Klein *et al.* 1986, Kurz 1986a, b, Nishiizumi *et al.* 1986, Phillips *et al.* 1986, Marti and Craig 1987, Nishiizumi *et al.* 1987). Since the publication of these pioneering studies, the methodological developments and their applications to Earth surface sciences has taken off impressively (Fig. 4.1). The technique has now clearly left the realm of specialist interest and has become a widely used tool in geomorphology. In fact, the novel ability to use cosmogenic nuclides to date geomorphologic surfaces, and determine process rates from rock, regolith and soils (Lal 1991), has revolutionized and rejuvenated many fields of geomorphology.

In this chapter, the main applications of cosmogenic nuclides will be visited in the following sequence: (i) exposure dating of geologic/ geomorphic surfaces, (ii) burial dating, (iv) erosion/denudation rate determinations, (vi) constraining uplift rates and (v) soil dynamics. The methodological basis for each will be provided, along with a discussion of both generic and specific examples. At the end of this chapter, avenues to deal with the inherent methodological and geologic uncertainties will be examined.

4.1 Exposure dating

In principle, any geological surface that is stable and continuously exposed to cosmic rays can be dated by measuring the amount of accumulated cosmogenic nuclide in surfacial rocks. It is therefore possible to constrain the timing

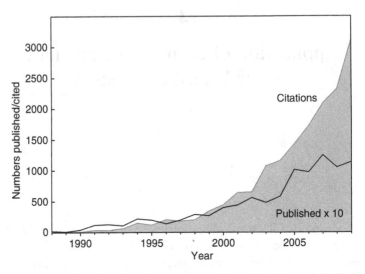

Fig. 4.1. Number of publications on terrestrial applications and citations thereof. The number of publications is multiplied by ten to have them visible on the common scale (Source: ISI Web of Knowledge, accessed 1 May 2009; values for 2009 are annualized; search terms: cosmogenic AND (exposure OR erosion OR denudation OR soil OR lava OR terrestrial OR in situ); NOT (meteorite OR chondrite OR lunar OR radiocarbon)).

of the land-forming process that created the surface, provided its duration was short relative to the age of the surface. These geological surfaces can be depositional features such as moraines, fluvial deposits, gravity and lava flows; erosional features carved by glacial, fluvial and aeolian forces; and features created by endogenic forces, such as fault scarps. The particularities of these settings will be discussed later in this chapter, after the general methodological framework for exposure dating has been set.

4.1.1 Non-eroding vs. eroding surfaces

Any surface that may be chosen for exposure dating can experience erosion during exposure – this can have profound effects on the eventual calculation of exposure ages and their meaning. For this reason, scenarios, both with and without erosion, are introduced consecutively in the following sections.

Non-eroding surfaces

A stable geologic surface exposed to cosmic rays will accumulate cosmogenic nuclides with exposure time t. If the surface is not eroding, the total cosmogenic radionuclide concentration C_{total}, at any subsurface depth z, can be described by:

$$C_{\text{total}}(t,z) = C_{\text{inh}}(z)e^{-t\lambda} + \sum_i \frac{P_i(z)}{\lambda}(1 - e^{-t\lambda}) \qquad (4.1)$$

(for derivations of this and most of the following equations see, e.g., Lal 1991, Niedermann 2002), with

$$P_i(z) = P_i(0)e^{-z\rho/\Lambda_i} \qquad (4.2)$$

(see also Section 1.4). C_{inh} denotes any component present at the start of exposure, which may be inherited from previous exposures and/or non-cosmogenic production (Chapter 2); the subscript 'i' denotes the different production pathways for any given cosmogenic nuclide (i.e. via spallogenic and thermal neutrons, muon capture, fast muons; see Section 1.4 and Chapter 2); ρ is the density of the sample's overburden and λ denotes the *decay constant*. Note that some practitioners prefer to use the *radioactive mean life* τ instead of λ, which is the reciprocal of λ, thus they are readily interchangeable. A radionuclide's *half-life* $T_{1/2}$ is the reciprocal of λ multiplied by the natural logarithm of two.

For short exposure histories ($t \ll T_{1/2}$) and for stable nuclides Eqn. (4.1) simplifies to:

$$C_{\text{total}}(t,z) = C_{\text{inh}}(z) + \sum_i P_i(z)t \qquad (4.3)$$

For cases where inheritance can be excluded ($C_{\text{inh}} = 0$), or determined independently, the cosmogenic nuclide concentration ($C_{\text{cos}} = C_{\text{total}} - C_{\text{inh}}$) can be used to calculate the *exposure age* T_{exp}:

$$T_{\text{exp}} = -\frac{1}{\lambda}\ln\left(1 - \frac{C_{\text{cos}}(z, T_{\text{exp}})\lambda}{\sum_i P_i(z) \cdot e^{-\rho z/\Lambda_i}}\right) \qquad (4.4)$$

(Lal 1991, Niedermann 2002). For surface samples ($z = 0$) this equation reduces to:

$$T_{\text{exp}} = -\frac{1}{\lambda}\ln\left(1 - \frac{C_{\text{cos}}(0, T_{\text{exp}})\lambda}{\sum_i P_i(0)}\right) \qquad (4.5)$$

As before, for short exposures ($t \ll T_{\frac{1}{2}}$) and stable nuclides, Eqns. (4.4) and (4.5) can be simplified to:

$$T_{\exp} = \frac{C_{\cos}(z, T_{\exp})}{\sum_i P_i(z) \cdot e^{-\rho z / \Lambda_i}} \qquad (4.6)$$

and

$$T_{\exp} = \frac{C_{\cos}(0, T_{\exp})}{\sum_i P_i(0)} \qquad (4.7)$$

respectively.

Eroding surfaces

In many cases, geological surfaces are modified by erosion during their exposure. For these cases Eqns. (4.1) and (4.3)–(4.7) are strictly no longer valid. Erosion will lower the measured cosmogenic nuclide concentrations in samples from such surfaces, by bringing material to the surface that had been previously (partially) shielded from cosmic rays. Because Eqns. (4.3)–(4.7) neglect erosion, using them to calculate exposure ages $T_{\exp}$ yields only *minimum ages*. If the amount of erosion can be determined/estimated independently, and a constant *erosion rate* ε is assumed, then Eqns. (4.4)–(4.7) can be used in conjunction with a correction factor. For cases with a relatively small amount of time-integrated erosion (<10 cm), and nuclide production dominated by spallation, an approximate erosion correction factor f_ε may take the form of:

$$f_\varepsilon = 1 + \frac{\varepsilon T_{\exp} \rho / \Lambda_{\mathrm{sp}}}{2} \qquad (4.8)$$

The exposure age (Eqns. (4.4)–(4.7)) corrected for erosion $T_{\exp,\varepsilon-\mathrm{corr}}$ would then be:

$$T_{\exp,\varepsilon-\mathrm{corr}} = f_\varepsilon T_{\exp} \qquad (4.9)$$

The accumulation of cosmogenic nuclides in eroding surfaces is accurately described by the general formula:

$$C_{\mathrm{total}}(t, z) = C_{\mathrm{inh}}(z) e^{-t\lambda} + \sum_i \frac{P_i(z)}{\lambda + \rho \varepsilon / \Lambda_i} e^{-\rho(z_0 - \varepsilon t)/\Lambda_i} \left(1 - e^{-(\lambda + \rho \varepsilon / \Lambda_i)t}\right) \,(4.10)$$

(Lal 1991, Niedermann 2002), which is valid for situations with a *constant erosion rate* ε; z_0 denotes the initial shielding depth ($z_0 = \varepsilon T_{\exp}$).

Naturally, this general formula is also valid for the situations where approximations of Eqns. (4.8) and (4.9) may be used. Analogous to Eqn. (4.5), for surface samples ($z_0 - \varepsilon\, T_{exp} = 0$) with negligible inheritance ($C_{inh} = 0$) and dominantly spallogenic production, exposure age can be calculated using:

$$T_{exp} = -\frac{1}{\lambda + \rho\varepsilon/\Lambda_{sp}} \cdot \ln\left(1 - \frac{C_{cos}(0, T_{exp})(\lambda + \rho\varepsilon/\Lambda_{sp})}{\sum_i P_i(0)}\right) \qquad (4.11)$$

If contributions by muogenic and/or thermal neutron production are significant, Eqn. (4.11) should not be used, and T_{exp} has to be derived numerically from Eqn. (4.10).

In the limiting case of long-term steady-state erosion (Lal 1991), the cosmogenic nuclide concentration is no longer time dependent, but is determined by the erosion rate; if $T_{exp} \gg 1/(\lambda + \rho\varepsilon/\Lambda_i)$, Eqn. (4.10) reduces for surface samples ($z_0 - \varepsilon\, T_{exp} = 0$) with negligible inheritance ($C_{inh} = 0$) to:

$$C_{total}(z) = \sum_i \frac{P_i(z)}{\lambda + \rho\varepsilon/\Lambda_i} \qquad (4.12)$$

(Lal 1991). In situations where steady-state conditions have been achieved by a given nuclide in a sample, no age information can be obtained, other than that the minimum time to reach steady state has elapsed (Fig. 4.2). This relationship (Eqn. (4.12)) is further explored in Section 4.3, where erosion and denudation rates are considered in detail.

In the general case (Eqn. (4.10)), the observed cosmogenic-nuclide concentration in a sample is a function of at least two independent variables: exposure age and erosion rate. Thus, if the erosion rate of a surface cannot be determined or estimated by independent methods, a single nuclide measurement can only provide a minimum exposure age (Eqn. (4.11)). In order to find a unique solution for both variables, the concentrations of at least two nuclides must be measured in the same sample.

In cases where a short-lived radionuclide, such as ^{14}C, has reached saturation (i.e. after $T_{exp} \sim 25$ ka for ^{14}C, see Fig. 4.2), its concentration can be used to derive the steady-state erosion rate (Eqn. (4.12); Section 4.3). This erosion rate can be used to derive the exposure age with Eqn. (4.10), using the concentration of long-lived [10]Be in the same sample (Gosse and Phillips 2001) and implicitly assuming that the average erosion rate determined for the saturation timescale (<25 ka) is valid

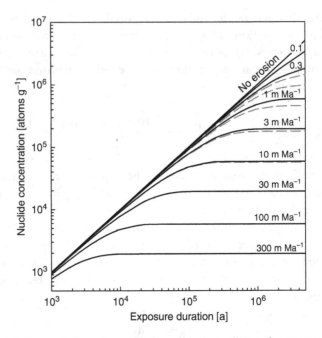

Fig. 4.2. Accumulation of cosmogenic nuclides in eroding surfaces; exposure time since onset of erosion of a surface with no pre-exposure memory. Black solid lines are for stable nuclides, grey stippled lines for ^{10}Be; both calculated for a hypothetical production rate of 1 atom per gram per year (multiply by actual local production rate). At low erosion rates the decay of ^{10}Be affects ^{10}Be concentrations; at higher erosion rates (>10 m/Ma) decay has no discernable effect on concentrations. Depending on the erosion rate $10^4 ->10^6$ years are required before constant nuclide concentrations reflect the steady state erosion rate; i.e. before secular equilibrium between production and decay/erosion is reached.

for longer timescales. In most sampling situations, ^{14}C is the only cosmogenic nuclide that may be saturated in surfaces, as other nuclides have much longer half-lives, and consequently take much longer to reach saturation (Fig. 2.1). However, in situ ^{14}C is not yet routinely analysed (Section 2.3.2), requiring alternative applications with two nuclides to solve the problem.

If the half-lives of two cosmogenic nuclides are sufficiently different, the ratio R_{AB} of their concentrations in a sample (Eqn. (4.13)) can be used to estimate the exposure age and erosion rate (Lal and Arnold 1985, Lal 1991, Gosse and Phillips 2001). It may also serve to identify and quantify the duration of intermittent burial (Section 4.2.2), and as a test of the steady-state erosion assumption. For surface samples, following from Eqn. (4.10), the ratio R_{AB} is defined by:

$$R_{AB}(0,t) = \frac{\sum_i \frac{P_{A,i}(0)}{\lambda + \rho\varepsilon/\Lambda_i}\left(1 - e^{-(\lambda + \rho\varepsilon/\Lambda_i)t}\right)}{\sum_i \frac{P_{B,i}(0)}{\lambda + \rho\varepsilon/\Lambda_i}\left(1 - e^{-(\lambda + \rho\varepsilon/\Lambda_i)t}\right)} \qquad (4.13)$$

The subscripts 'A' and 'B' may denote any cosmogenic radionuclide. ^{10}Be and ^{26}Al are the most commonly used pair (Fig. 4.3). This is because both can be readily analysed in quartz (Section 2.3). However, because their half-lives differ only by a factor of two (Section 2.3), the ratio R_{AB} is not a very sensitive function of erosion rate and exposure duration. Thus, the 'island of steady-state erosion' (Lal 1991) is relatively narrow (Fig. 4.3). Analytical uncertainties for ^{26}Al are commonly much larger than those for ^{10}Be (Section 2.3) and often preclude an accurate resolution of erosion rate and exposure age. Combination of e.g. ^{36}Cl and ^{10}Be from minerals in the same rock has the potential to resolve age and erosion rate ambiguities in eroding surfaces (Ivy-Ochs *et al.* 2007). For old surfaces, the combination of stable nuclides with radionuclides is promising, e.g. the pair ^{21}Ne and ^{10}Be in quartz has been used successfully (Kober *et al.* 2007).

For situations where deep (≥ 1.5 m) depth profiles can be sampled, yet another approach is feasible to constrain exposure ages and erosion rates. In these cases the concentration vs. depth profile of a single nuclide can be matched to model predictions (i.e. depth profiles calculated using a physical model incorporating multiple pathways such as spallogenic and muogenic production), and the most likely erosion rates and exposure ages be constrained e.g. using chi-squared minimization (Braucher *et al.* 2009, Hein *et al.* 2009, Schaller *et al.* 2009). In practice, depending on the analytical uncertainties for samples at individual depths, at least six samples, over a depth interval in the order of 1.5 m or more, are required to achieve good results (Braucher *et al.* 2009, Hein *et al.* 2009, Schaller *et al.* 2009; Fig. 4.4).

4.1.2 Mass shielding

Any solid mass that is sufficiently thick to block or attenuate the cosmic-ray flux at or near the sampling site, will affect the cosmic-ray flux – and the associated cosmogenic-nuclide production – at the site. To obtain accurate ages, the surface production rates P_0, as used in the equations

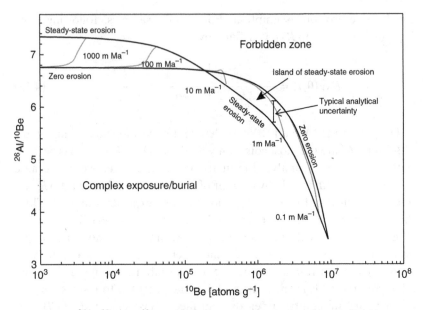

Fig. 4.3. ^{26}Al/^{10}Be vs. ^{10}Be diagram, calculated including the contributions of muons. At high erosion rates nuclides produced at great depth by muons can reach the surface before they decay, thereby increasing the relative contribution of ^{10}Be and raising the ^{10}Be/^{26}Al ratio above that for a stable surface. The general shape of the curves is due to the different half-lives of ^{10}Be and ^{26}Al, i.e. ^{26}Al decays faster than ^{10}Be. After > 8 Ma exposure both nuclides reach saturation and the plot ends at an invariant point. Samples from stable surfaces with no erosion will move along the 'zero erosion' line in the direction of higher ^{10}Be concentrations; samples from steadily eroding surfaces stay put at a certain point on the 'steady state erosion' line. The grey lines depict model trajectories for eroding samples on their way to steady state (Fig. 4.2). The end-point of the grey lines is where samples will remain if the corresponding steady state is reached. The steady state erosion line links all possible steady state solutions. All samples with a simple single exposure and erosion history will lie between the two black lines, on the 'island of steady state erosion' (Lal 1991). Samples that have a complex exposure history, which includes periods of burial, lie below the 'island'. The area above and to the right of the island is forbidden, i.e. no exposure scenario can produce results in that region. The indicated typical uncertainty of ^{26}Al/^{10}Be ratios is calculated assuming ±5% (±1σ) uncertainty for ^{10}Be and ^{26}Al determinations. After Lal (1991).

introduced in Section 4.1.1, need to be corrected (reduced) accordingly by multiplication with correction factors *f*. Corrections for topographic shielding, self-shielding and shielding by overburden, such as snow and soil, are commonly applied, and are discussed in the following.

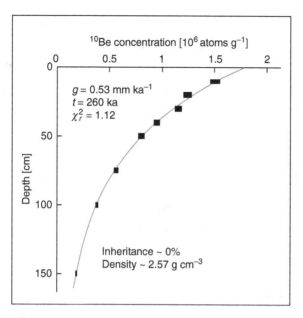

Fig. 4.4. [10]Be depth profile of amalgamated pebble samples from a glacial outwash in Patagonia. The size of the black rectangles represents the uncertainties (±1σ) of sample depth and [10]Be-concentrations. The results lie on a forward modelled profile (grey line); which is calculated using age-constraints from surface samples and the sediment density as external constraints. The good fit of measured and modelled data constrains the erosion rate of the outwash gravel plain, and further precludes significant pre-exposure and cryoturbation below the highest sampled depth-level. After Hein *et al.* (2009).

Topographic shielding

The maximum cosmic-ray flux that can be received by a sample at a given geographic location is on a flat, horizontal surface with an unobstructed view of the sky. Any obstruction sufficiently thick to block cosmic rays will lower both this flux, and the corresponding production rate. Therefore, one must account for the shielding of the sampling site from cosmic rays by the surrounding topography in order to obtain accurate exposure ages. Fortunately the cosmic-ray flux is strongly biased to the vertical (see Section 1.3), and thus corrections are usually relatively small (Fig. 4.5).

For rectangular obstructions – obstructions that define the cosmic-ray horizon (usually equivalent to the visible horizon) – with a constant inclination θ above the horizontal and extending over an azimuthal width

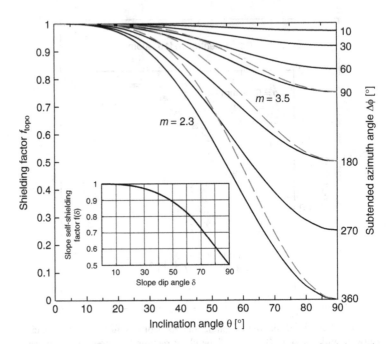

Fig. 4.5. The topographic shielding factor f_{topo} as a function of height and lateral extent of a rectangular obstruction; the height being expressed as inclination above the horizon, the lateral extent as subtended azimuth angle, both as seen from the sampling location. The black lines are calculated assuming $m = 2.3$ for the exponent in the equation describing the angular dependency of cosmic ray flux (Eqn. 1.4); the grey stippled lines are calculated for an alternative value of $m = 3.5$ (shown for 90°, 180° and 360° subtended azimuth angle, for discussion on values of m see Section 1.3). The inset shows the self-shielding of a dipping surface ($m = 2.3$). For illustration: if the entire horizon is limited by obstructions visible under 20° inclination the cosmic ray flux decreases by 3% ($m = 2.3$); if the continuous obstruction is 40° high the flux is decreased by 23% ($m = 2.3$). The corresponding values for $m = 3.5$ are 1% and 14%, respectively. The self-shielding of surfaces dipping at angles shallower than 30° is <3% (see inset). Figure after Dunne (1999).

of $\Delta\varphi$, the fraction of the remaining cosmic-ray flux relative to the unobstructed flux is given by

$$f_{\text{topo}} = 1 - \frac{\Delta\varphi}{2\pi}\sin^{m+1}\theta \qquad (4.14)$$

(Dunne *et al.* 1999; Fig. 4.5).

For a series of n obstructions, each visible from the sampling site under an individual θ_i, and $\Delta\varphi_i$ wide, this fraction is

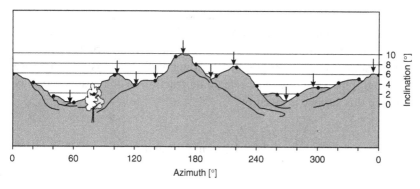

Fig. 4.6. To assess the topographic shielding at a sampling site the inclination of the visible horizon is measured using an inclinometer. If the horizon
is obscured by trees the location of the horizon should be estimated by
interpolation. Various approaches are used by practitioners; some prefer
readings at constant intervals (black circles, here at 20° intervals), others
choose fewer points that allow the approximate description of the horizon
by straight lines (black arrows). The differences between approaches are
marginal, and the results are essentially identical. The depicted situation
yields shielding factors of 0.99957 and 0.99961, for horizon definitions using
the black circles or the arrow end-points to define the horizon, respectively
($m = 2.3$). To illustrate that topographic shielding is only significant in rather
'spectacular' landscapes: if the topography were twice as steep as depicted
(i.e. multiply the inclination values by two) the topographic shielding factor
would be 0.996; corresponding three-fold and four-fold steepening would
result in shielding factors of 0.986 and 0.965, respectively.

$$f_{\text{topo}} = 1 - \frac{1}{360°} \sum_{i=1}^{n} \Delta\phi_i \sin^{m+1} \theta_i \qquad (4.15)$$

(Dunne *et al.* 1999). For triangular obstructions, with a baseline $\Delta\phi_i$ wide,
θ_i can be replaced by

$$\theta_{R,i} = 0.62\theta_T - 0.00065\theta_T \qquad (4.16)$$

(Dunne *et al.* 1999), with θ_T being the maximum inclination angle (in
degrees). The above equations represent a standard approach to correct
for the topographic shielding of a sampling site, as shown in Fig. 4.6.

A commonly used value for the exponent m is 2.3, however, other
exponents as high as 3.5 have been used (for references and discussion
see Section 1.3). In practice, for most sampling situations, the absolute
differences that arise from the choice of m are small (Fig. 4.5). These
differences are only significant in cases of extreme shielding (i.e. cliff-faces
or oversteepened canyons).

Self shielding

If the sampled surface dips at an angle δ, and has a lateral extension of several metres or more, part of the cosmic-ray flux will be blocked in a manner similar to the topographic shielding; in fact, this is a special case of topographic shielding. The remaining fraction of the cosmic-ray flux reaching this surface is given by:

$$f(\delta) = \frac{m+1}{2\pi} \int_{\phi}^{2\pi} \int_{\theta=\gamma(\phi,\delta)}^{\pi/2} \sin^m\theta \, \cos\theta \, d\theta \, d\phi \qquad (4.17)$$

(Dunne *et al.* 1999, Niedermann 2002), with the slope angle γ in direction φ given as

$$\gamma(\varphi,\delta) = \arctan(\cos\varphi \tan\delta) \qquad (4.18)$$

(Niedermann 2002). The relation between surface dip angle and relative cosmogenic production rates is shown in Figure 4.5. Equation (4.17) can also be used for situations where the inclined surface is dipping in the direction of further topographic shielding; i.e. in situations with self- and additional topographic shielding (Dunne *et al.* 1999).

In practice, sampled surfaces are rarely part of a large planar surface extending laterally for several metres. In these cases a 'normal' topographic shielding correction is sufficient, particularly when the surface is inclined by less than 30° (Balco *et al.* 2008). Special cases of extreme shielding, and the effect of shielding on attenuation coefficients relevant for subsurface samples (depth profiles) are discussed in Gosse and Phillips (2001) and Dunne *et al.* (1999).

Another form of self-shielding is the result of the physical thickness of a sample, as it is impossible to collect infinitely thin samples. The bottom of a sample is shielded by the mass of the entire sample; samples closer to the surface are correspondingly less shielded. For spallogenic reactions (Section 1.5) in moderately thick samples (<10 cm) the mean fraction of the remaining cosmic-ray flux can be approximated by:

$$f(z) = e^{-z\rho/2\Lambda_{sp}} \qquad (4.19)$$

after Eqn. (1.5), with z being the mean thickness of the sample, ρ its density and Λ_{sp} the attenuation coefficient for spallation reactions (Section 1.4). Reactions with significant cross-sections for thermal and epithermal neutrons that vary in a non-trivial fashion with depth (Fig. 1.6) require more elaborate correction factors (Gosse and Phillips 2001). Generally, to minimize uncertainties associated with the depth correction, it is prudent to collect relatively thin samples (<5 cm), if practically possible.

Other mass shielding/surface coverage

Any temporal or semipermanent overburden covering the site, such as snow, peat, soil, loess, volcanic ash or vegetation, will also reduce cosmic-ray flux and production rates. For spallogenic reactions, the corresponding correction factor has the general form

$$f_{cover} = e^{-z\rho/\Lambda} \tag{4.20}$$

after Eqn. (1.5), with z denoting the thickness of the cover, ρ the cover's density and Λ the attenuation coefficient for spallation reactions (Section 1.4). Note that the density of cover material may be highly variable and needs to be determined (e.g. loess, soil or ash) or estimated (snow, peat). The (variable) water contents of, for example, peat or soil can change their densities significantly.

The snow-cover correction is the most commonly applied correction. It may take the form of:

$$f_{snow} = \frac{1}{12} \sum_{i}^{12} e^{-(z_{snow} - z_{sample})\rho_{snow}/\Lambda} \tag{4.21}$$

(Gosse and Phillips 2001), where z_{snow} is the mean monthly snow depth, and z_{sample} is the height of the sample surface above the surrounding ground (e.g. in the case of a boulder). The format of Eqn. (4.21) allows accounting for the annual variability in snow cover. Snow density is highly variable, e.g. a density range 0.16–0.33 g cm^{-3} appears to be typical for northern Eurasia (Onuchin and Burenina 1996). To illustrate the potential magnitude of a snow-cover correction: if snow cover persists for ~4 months per year, it needs to be between 75–150 cm thick in order to reduce the annual cosmogenic nuclide production by 5% (using the above density range).

When performing a snow-cover correction, it is useful to consider how representative the (short/recent) snow-cover record is for the duration of the (long) exposure history investigated. If it cannot be established with reasonable confidence that the short-term cover record may be representative for the long term, an *elaborate* cover correction may be pointless; as such a correction may accurately describe a situation that has never occurred. In ambiguous situations, a correction may be replaced by a discussion of the robustness of the data in the light of possible significant, but loosely constrained, snow cover.

In order to avoid uncertainties associated with snow-cover correction it is often attempted to sample surfaces that are unlikely to be covered by a significant amount of snow. For instance, glacial boulders taller than >2 m may be the preferred target in regions with significant snow cover (John Gosse, personal communication).

Vegetation shielding is usually not significant, but may reduce the cosmic-ray flux by ∼2–7% in particularly dense forests (Cerling and Craig 1994a, Plug *et al.* 2007). For vegetation corrections, the average vegetation mass is, conceptually, evenly spread over the sampling area (Eqn. (4.20)). In areas with very large trees (temperate or tropical rainforests), the vegetation correction may be spatially variable for short timescales (Plug *et al.* 2007).

For slowly evolving (increasing/decreasing) overburden, as may be the case for loess, soil and ash cover, a plausible age-evolution model needs to be devised in order to make a suitable cover correction. If this is not possible, then, as discussed for snow-cover correction, a correction may be replaced by a discussion of the robustness of the data in the light of possible significant, but loosely constrained, overburden.

Most forms of overburden shielding may have significant water content (or are essentially water in the case of snow). Therefore, the thermal-neutron flux at the overburden/rock interface may be greatly affected by the overburden (Section 1.4). However, there is currently no suitable formulation to account for these interface effects (Phillips *et al.* 2001).

Uplift/subsidence

Uplift and subsidence corrections are special forms of the overburden correction. Uplift 'removes' part of the atmospheric overburden, while subsidence of a sampling site increases atmospheric mass-shielding (Section 1.3). Equation (4.20) may be used for this correction, using the appropriate air densities for the atmospheric depth interval the sampling site has traversed and an appropriate attenuation coefficient for the atmosphere (Section 1.3). Alternatively the vertical movement can be considered when calculating scaling factors (Chapter 3) for the site (which is mathematically equivalent). This correction can be significant for old sites with considerable time-integrated vertical movements (Brook *et al.* 1995a, Schäfer *et al.* 1999, Dunai *et al.* 2005), and is discussed as a potential application in Section 4.4.

4.1.3 Dating of depositional surfaces

Surfaces on sedimentary deposits formed by glacial, fluvial and marine activity are common targets for exposure dating (cf. Gosse and Phillips 2001; Putkonen and Swanson 2003). Debris from rock falls and other gravitational deposits have also been dated successfully (Kubik *et al.* 1998, Hermanns *et al.* 2001). Due to the nature of the processes delivering the sediments, and the often unconsolidated nature of the sediments themselves, inheritance from previous exposure, and postdepositional erosion and perturbation, are particularly important considerations for exposure dating of such deposits.

A special case of depositional surfaces are volcanic deposits. Features relevant for exposure dating of lava flows are discussed at the end of this section.

Pre-exposure

Debris in glacial deposits may have had a significant pre-exposure before erosion and transport by the glacier, either at or near the surface of the scoured bedrock, or at or near the surface of cannibalized older glacial sediments. Similar considerations hold for fluvial sediments that may have accumulated cosmogenic nuclides during initial erosion and/or protracted transport to the sampling site. Gravitational mass-wasting deposits may contain a random mixture of material previously exposed at/near the surface of the failing hill slope, and formerly deeply buried material.

Due to the random nature of erosion, transport and deposition, the specific pre-exposure of an individual boulder, pebble or sand grain usually cannot be predicted with confidence. The effects of pre-exposure may, however, be estimated for representative multigrain/multipebble samples (Anderson *et al.* 1996, Repka *et al.* 1997) or a representative group of results from individual cobbles and boulders (Rinterknecht *et al.* 2006).

Anderson *et al.* (1996) proposed assessing a deposit's inheritance by amalgamating 30–50 or more clasts or grains into a single sample before analysis. If this is representative of the average inheritance of a deposit, then an amalgamated sample from the surface will represent the total concentration (Eqns. (4.1), (4.3) and (4.10)) of cosmogenic nuclides accumulated prior to and after deposition. An equivalent amalgamated sample (same grain size, same lithology) from depth (>2–3 m) will only contain the inherited component (Fig. 4.3). The exposure age is calculated from the difference between the two concentrations (Anderson *et al.* 1996,

Repka *et al.* 1997, Hancock *et al.* 1999, van der Wateren and Dunai 2001). Alternatively, a detailed depth profile can establish the average pre-exposure (Fig. 4.3). This paired amalgamated sample approach is strictly only valid if the surface samples and the depth sample belong to the same sedimentary unit, i.e. were deposited in the same cycle/event. The sediments within one unit may be assumed to have had the same average erosion/transport history; i.e. that representative sediment samples will have a uniform pre-exposure.

The deep sample may be replaced/supplemented by an equivalent modern sediment sample from an active channel cutting into its own sediment terraces (i.e. those dated), though only if it can be assumed that the modern sediment delivery occurs via similar processes, at a similar rate as the sediments on abandoned levels (Hetzel *et al.* 2002b, González *et al.* 2006). This may be assumed when the timescales of abandonment are short compared to long-term climatic changes.

If a suitably large data set describing age-equivalent deposits is available, and inheritance is a relatively rare and random occurrence, ages of individual cobbles and boulders that are affected by inheritance can be identified as outliers and excluded from further consideration (Rinterknecht *et al.* 2006; see also Section 4.6). Inheritance on glacial moraines is apparently rare; from a recent compilation of published moraine data it was concluded that less than 3% of boulders on glacial moraines appear to have pre-exposure (Putkonen and Swanson 2003).

Erosion and perturbation of moraines

Of the depositional surfaces typically dated using cosmogenic nuclides, moraines are particularly sensitive to erosion. In essence, they are often a steep pile of sediments, originally starting off as sharp-crested features close to the angle of repose (Putkonen and Swanson 2003). Consequently, their local erosion potential and gradient are at, or close to, the maximum that can be achieved for unconsolidated sediments. The fine-grained matrix is easily dislocated by fluvial and aeolian erosion, as well as gravitational creep. Thus, over time, boulders will emerge from their original (partially) shielded position in the moraine and join the previously/continuously exposed boulders on the degrading moraine ridge. Upon exposure, boulders invariably weather and erode at rates which are a function of local climate and boulder lithology (cf. Putkonen and Swanson 2003; Fig. 4.7).

Due to their precarious nature, originally sharp-crested moraines relax relatively quickly by diffusive grain transport, broadening and reducing in

Fig. 4.7. Emerging granite boulder on a moraine in Patagonia, Argentina. Lichen growth and weathering contrasts indicate that the lower third of the exposed part of the boulder was relatively recently exposed by exhumation. The sharp line between weathered and unweathered parts of the boulder suggests exhumation occurred during a brief episode, rather than by continuous erosion of the moraine. The boulder is positioned on the flank of the moraine, i.e. a position that is particularly sensitive to boulder exhumation by erosion; this boulder was *not* sampled for exposure dating.

height (Putkonen and Swanson 2003). The amount of lowering is a function of initial height, age of the moraine and topographic diffusivity (Putkonen and Swanson 2003). Under constant environmental conditions, the lowering rate decreases with the moraine size (Putkonen and Swanson 2003). An important feature is that moraines initially degrade quickly, i.e. most boulders, if any, are exposed soon after the glacier's retreat (Putkonen and Swanson 2003). Thus, depending on the age and size of the moraine, it appears, from modelling results, that of 3–7 randomly selected boulders from the moraine crest at least one (the oldest) will show an age that is within 10% of the moraine age (with 95% confidence; Putkonen and Swanson 2003). This model prediction might not apply to areas with strong wind erosion, as this situation is actually not within the bounds of the model, which is reliant on diffusive processes. Recent findings appear to corroborate this limitation (Hein *et al.* 2009).

Moraine boulders should ideally be sampled on the ridges only, not on steep flanks. In a flank position, exhumation potential is maximal. Moreover, boulders may be gravitationally rotated and positioned in their present orientation at an unspecified time in the past.

In several favourable settings, moraine erosion may be negligible. Holocene moraines may be almost perfectly preserved, due to insufficient time for significant erosion to have taken place. In areas with extremely low erosion rates, such as found in certain regions of Antarctica, old moraines may also remain well preserved (Brook *et al.* 1993). Very small moraines may lower less than the diameter of the smallest boulder sampled (Putkonen and Swanson 2003). Finally, not all moraines start off sharp crested; composite moraines may start off relatively level, and remain level due to lack of local potential and gradient. In such situations moraine degradation will remain small (Putkonen and Swanson 2003).

In areas with human activity, removal of boulders from fields (Graf *et al.* 2007) and levelling of moraines by ploughing can be obstacles for exposure dating. If moraines are/were covered by tree growth, it should also be kept in mind that sizeable boulders can also be exposed or turned by tree-throw.

Boulder erosion will also affect exposure ages (Section 4.1.1) and can be addressed with various approaches of variable accuracy. The accuracy of such corrections tends to decrease with increasing cumulative erosion. If boulder erosion is small, glacial striations are still visible and/or fluvioglacial surfaces are still smooth (Fig. 4.8), erosion can accurately be constrained as being negligible. If boulders contain weathering-resistant quartz veins, their emergence from a formerly smooth surface can be used to assess and correct for the time-integrated erosion rate (Balco *et al.* 2009). If these features are absent, the degree of edge-rounding or preservation of fluvioglacial shapes can be used as qualitative guidance for the time-integrated erosion on the centimetre to decimetre scale (Hein *et al.* 2009; Fig. 4.9). If erosion obliterated all of the above features, the application of a suitable pair of cosmogenic nuclides (Section 4.1.1) may help to constrain the erosion rate. A note of caution: boulders that appear remarkably unweathered when compared to their peers of the same lithology on the same moraine may be recently exhumed. If there is no lithological explanation for such exceptionally good preservation (e.g. vein-quartz boulders are particularly weathering- and erosion-resistant), such boulders should probably be avoided for exposure dating.

Fig. 4.8. Glacial polish preserved on relict patches on an andesitic moraine boulder in Peru (Altiplano). The smooth glacially polished relicts show as shining patches on the boulder. From the surface roughness and the elevation of the relicts above the weathered portions the cumulative erosion of the weathered portions can be estimated to be less than 2 cm (hammer for scale).

Erosion and perturbation of fluvial and beach sediments

Fluvial sediments on floodplains commonly form gently dipping to near-horizontal surfaces. Once such a level surface is abandoned by incision of the fluvial system that originally deposited the sediment, it is difficult to move material laterally on the emergent terraces. In the centre of the terraces, away from the terrace margins, local gravitational potential and gradients are near the minimum. Sand and finer particles may be transported by overland flows, or blown out by wind, yet pebbles and cobbles are, in many cases, essentially immobile due to the lack of a suitable, high-energy transport mechanism. Over time, sediment terraces erode back and a shallow drainage system can develop. Radiating from the terrace edges and the newly developed terrace drainage, diffusive processes induce surface rounding, eventually mobilizing formerly immobile coarse sediments, which persist longest on the interfluves of the emergent terrace drainage.

Fig. 4.9. Wind-blasted andesitic moraine boulder in Patagonia, Argentina. The top and the left side of the boulder still preserve the original fluvio-glacial shape of the boulder, the top (hammer for scale) with glacial polish. The wind-ward right side (the western), and the front, are severely eroded by wind (fluting, Viles and Burke 2007). Assuming a regular fluvio-glacial shape, it can be estimated that these sides lost several decimetres by wind-erosion.

Once this state is reached, i.e. when the surface rounding of the terrace landform is pervasive, it is difficult to use exposure dating to constrain the terrace age. More steeply inclined sediment surfaces, such as may be found on alluvial fans (Siame *et al.* 1997, Cerling *et al.* 1999, Daeron *et al.* 2004, Nishiizumi *et al.* 2005, González *et al.* 2006, Duhnforth *et al.* 2007, Frankel *et al.* 2007, Carizzo *et al.* 2008, Evenstar *et al.* 2009) or on abandoned beach levels (Perg *et al.* 2001, Matmon *et al.* 2003, Owen *et al.* 2007, Quezada *et al.* 2007) go, in principle, through the same stages of landform degradation as described above.

How quickly a sediment surface is 'spoiled' for exposure dating is dependent on many factors, such as the nature of the sediment (coarse grained, fine grained, sorted/ unsorted etc.) and the availability of water. Induration of sediments by calcrete or silcrete in semi-arid landscapes, or stabilization by a continuous vegetation cover in a moderate climate, can greatly reduce topographic diffusivity and help to retain original planar sections for long periods (Hancock *et al.* 1999, van der Wateren and Dunai 2001). Progressive terrace incision can isolate mesa-like terrace remnants, which are subsequently protected from flash floods in arid environments (Dunai *et al.* 2005). In short, many sediment surfaces may have a remarkable resilience to erosion. Generally, due to their lower gradients, they may have a much greater preservation potential than moraines; for instance, outwash plains may be better preserved than associated moraines (Hein *et al.* 2009).

The preservation of original planar landforms is a good first indicator of the stability of a sedimentary terrace. Their preservation may be established from field mapping, aerial photography, differential GPS surveying (González *et al.* 2006), levelling (Hetzel *et al.* 2002b) or LIDAR.

One disadvantage of fluvial sediments, from the dating perspective, is that abandoned river terraces are often prized agricultural land. Cobbles are often removed from fields, and the uppermost several decimetres are disturbed by ploughing, making the dating effort more challenging, if not impossible. In addition, bioturbation can thoroughly mix several decimetres of surface material over time. By sampling suitable depth profiles, surface mixing can be detected and corrected for (Perg *et al.* 2001, Schaller *et al.* 2009).

In periglacial environments, cryoturbation can thoroughly mix glacial deposits and fluvial sediments. Again, depth profiles can be used to identify and correct for cryoturbation (Fig. 4.3).

Under favourable conditions, visible erosion contrasts on pebbles and cobbles can provide valuable information on the stability of a sediment surface. In regions with past/current aeolian activity, ventifaction occurring uniformly only on the top side of surface clasts indicates that the clasts have not been moved laterally (otherwise some would have turned over and have ventifacts on the bottom), and that clasts were exposed to aeolian activity for a significant period of time (van der Wateren and Dunai 2001; Fig. 4.10). Only under exceptional circumstances – such on the Racetrack Playa, Death Valley (Reid *et al.* 1995) – is significant lateral transport of clasts possible, without turning clasts.

In areas affected by cryoturbation, some clasts may have ventifacts on various/all sides, without significant lateral transport, rather indicating a long surface residence time (Hein *et al.* 2009). Of common geological

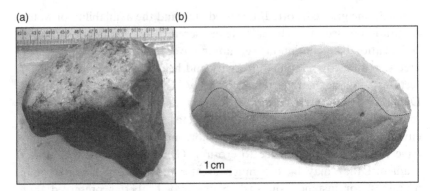

Fig. 4.10. (a) Cobble-sized vein-quartz from an alluvial fan in the Ugab catchment in Namibia. The dark, fully angular portion of the clast was in the soil. The emergent (white) portion of the clast is fully wind-polished. This erosion contrast indicates that the clast was in its sampled orientation for its entire exposure at the surface; i.e. lateral transport can be excluded (see text for discussion). A [10]Be exposure age of 460 ka was determined for this clast (concordant ages were obtained from equivalent clasts from the same surface). (b) Cobble-sized vein quartz from a fluvial deposit in the Ugab catchment in Namibia (different location than A). The fluvially rounded bottom of the clast is partially encrusted with pedogenic carbonate, the top is facetted by ventifaction. The thin black line denotes the boundary between fluvially and aeolian shaped portions of the clast. The absence of ventifacts on the bottom of the clast indicates that the clast was in its sampled orientation for its entire exposure at the surface.

materials, vein-quartz clasts are the most resistant to aeolian erosion and will withstand its attack longest.

As a result of continued aeolian activity, sediment surfaces containing pebble- to cobble-sized material may form lag deposits. In such cases, by comparing the clast density on the surface to the corresponding clast density in the underlying sediment, the amount of deflation of fine sediment can be determined (Hein *et al.* 2009). In the case of inflationary desert pavements (Wells *et al.* 1995) this approach is obviously not applicable.

Lithologies that are susceptible to weathering may develop a weathering contrast, which is analogous to the aeolian case discussed above, and can be used to exclude pervasive lateral transport of clasts. Depending on the availability of weathering agents, either the tops or the bottoms of clasts may be more weathered (e.g. in arid regions, the underside of clasts may be more weathered than the top, due to the availability of soil moisture and salt; the situation may be reversed in wet climates). Also, in terms of weathering, vein-quartz clasts are the most resistant; if available, they are an ideal sampling target for exposure dating of sediment surfaces.

Even if several, or all, of the above indicators point in the direction of a long-term stable surface, significant erosion might have affected a sediment surface targeted for exposure dating. For instance, lag deposits may be in steady-state equilibrium between aeolian erosion and replenishment from subsurface material. In such cases, exposure ages of individual clasts should vary widely, due to their different exhumation and exposure trajectories. The absence of a significant scatter of individual ages can therefore be used as an indication for surface stability. If pre-exposure can be excluded (see above), the magnitude of scatter of individual ages should be in accordance with the amount of time-integrated erosion (Hein *et al.* 2009).

Exposure dating of lava flows

Pre-exposure is not an issue for lava-flow tops; in most cases erosion and intermittent soil and ash cover are the main concerns. Inheritance of ^{10}Be from subduction recycling should be considered as a possibility for lavas related to arc-volcanism (Morris *et al.* 1990), and investigated using shielded samples (see above; analogous to investigating pre-exposure in sediments). Most lava flows are dated using cosmogenic ^{3}He and ^{36}Cl, which are not recycled by subduction.

The identification of erosion on a lava-flow top requires some knowledge of the appearance of a corresponding fresh, uneroded lava. The development of flow-top features is strongly dependent on the chemistry of the lava, as lava viscosity increases significantly with increasing silica content.

Acidic (i.e. silica rich) lava flows of rhyolitic composition are relatively rare. Often, features that may appear to be a rhyolite flow today are, in fact, ignimbrites (welded tuff), which may exhibit features like columnar jointing, air bubbles and flow textures. At the time of deposition, ignimbrites are invariably covered by unwelded tuff. The welding of ignimbrite occurs at a certain depth (dependent on eruption temperature, chemistry of the tuff and weight of overburden), aided by the pressure of the overburden of unwelded tuff. Subsequent erosion may remove unwelded tuffs and leave the erosion-resistant ignimbrite behind. When and how fast this uncovering occurs is a priori unknown. Exposure dating of ignimbrite surfaces will therefore always yield minimum ages only (Goethals *et al.* 2009a, Goethals *et al.* 2009b).

Fig. 4.11. Block lava flow, Atacama, Chile. The sharp-edged fully angular shape of the blocks and the absence of spalling fragments indicates a marginal erosion (<5 mm). The ^{3}He exposure age of this flow is ∼100 ka.

Intermediate volcanic rocks (e.g. andesites) may form streams of block lava. Such lava flows are commonly covered by blocks of regular form, often approaching a cube-like form, and with smooth surfaces. Block lava streams are formed by breaking of the partially to fully congealed upper part of a moving flow. Block flows can have a considerable thickness of up to a few tens of metres. The blocky material usually constitutes a substantial part of the flow, sometimes the entire flow (Fig. 4.11). The surfaces of block lava flows are very irregular. The preservation of block angularity can be used as a criterion to assess the presence of significant erosion and to estimate its total amount. Assuming a diffusive weathering process, edges of angular blocks of isotropic material recede faster than the planar surfaces of the same block (Shim 2002). The amount of edge recession can, therefore, be used as a conservative upper limit for surface erosion of angular boulders.

Basaltic lavas have the lowest viscosity of natural silicate melts. They can form pahoehoe flows that are characterized by smooth, billowing, undulating or ropy surfaces. Details of surface features of

Fig. 4.12. Exhumed ropey flow-top features of a pahoehoe lava on the uplifted flank of a pressure ridge, Etna, Sicily (Pete Burnard for scale). Remnants of the former covering flow are visible in the step-like feature above the ropey surface. This picture is to illustrate that only laterally extensive pahoehoe structures may be indicative for a continuous exposure. This would *not* be a suitable sampling site (if the age of the flow is the target).

pahoehoe lavas are typically on the centimetre scale. Their preservation can be taken as an indicator of low erosion rates (cumulative erosion is smaller than the smallest preserved original surface feature). While pahoehoe flow-top features are typically present at the surface, they can also form in lava tubes or survive covering by subsequent flows (Fig. 4.12). The presence of a well-preserved fragment of pahoehoe lava is therefore not a guarantee for the absence of erosion. Laterally extensive preservation of continuous pahoehoe features (i.e over several metres) and the absence of step-like features, however, can usually be taken as a reliable indication that the original lava surface is exposed, rather than it being a deeper level exposed by erosion.

Hornitos and pressure ridges (Fig. 4.12) on pahoehoe lavas are protruding features that are protected from persistent soil, ash and snow cover and are therefore valuable sampling targets for exposure dating (Cerling and Craig 1994a, Licciardi *et al.* 1999, Licciardi *et al.* 2008).

On cooling, degassing and/or partial crystallization, basaltic melts also become more viscous, leading to the emergence of aa lavas. These

Fig. 4.13. Rubbly surface of an aa-lava flow in Fuerteventura. Fragments of pahoehoe lava are visible in the foreground (left).

lavas are characterized by a rough or rubbly surface composed of broken lava blocks (Fig. 4.13), which are often cinder-like. The basaltic lava blocks are often more irregular than blocks of corresponding andesitic lava (see above). Aa flows may also be made up from fragments of pahoehoe lava, thus there is a continuum of surface features between these two end members. In order to judge the preservation of aa-lava blocks of unknown age it is useful to investigate small-scale surface features on equivalent material from recent/historic lava flows, if they are available in the vicinity. The preservation or absence of indicative surface features can be used to estimate the cumulative erosion. The approach to assess the preservation of old lava flows of unknown age via comparison to modern/historic equivalents in the same volcanic province is, of course, also a useful strategy for all other lava types.

Intermittent ash cover is a potential issue for most lava flows and should be investigated and excluded, if possible. Tightly clustering exposure ages from several samples collected over a large area on the same flow, and/or from features of variable height above the average surface of the lava flow, indicate that intermittent ash cover is probably not an issue at a particular site (ash cover should affect the different sampling points differently, and thus introduce scatter, if significant; Dunai and Wijbrans 2000). When sampling a wider area on the same flow, care has to be taken that sample collection is not extended into a neighbouring flow of similar surface expression, but different age (Dunai 2001a).

4.1.4 Dating of erosional surfaces

Any process that deeply excavates bedrock, i.e. exposes previously unexposed rocks at the surface, can, in principle, be dated using cosmogenic nuclides. Potentially suitable features encompass, for example, fluvially cut bedrock (Burbank *et al.* 1996, van der Wateren and Dunai 2001, Schaller *et al.* 2005), wave-cut platforms (Stone *et al.* 1996b, Lifton *et al.* 2001), glacially striated bedrock (Nishiizumi *et al.* 1989, Davis *et al.* 1999, Li *et al.* 2008), rock-fall scars and ancient human excavations (Cerling and Craig 1994b).

Tectonically exhumed fault scarps are a special type of erosional surface (*sensu lato*) (Mitchell *et al.* 2001, Benedetti *et al.* 2002, Benedetti *et al.* 2003, Palumbo *et al.* 2004), which are discussed at the end of this section.

As for the depositional surfaces discussed in Section 4.1.3, it is important that erosion can be excluded or quantified via the (partial) preservation of indicative original surface features. Pre-exposure, particularly by the deeply penetrating muogenic component, needs consideration in most cases. In the following sections, specific issues for various types of erosional surfaces are discussed.

Glacially polished bedrock

Glacially polished and striated bedrock surfaces were amongst the first sampling targets for exposure dating (Nishiizumi *et al.* 1989), not least because the presence or absence of erosion can be assessed by the (partial) preservation of glacial striations. In recent years, however, bedrock surfaces have lost some favour for studies of glacial chronologies as it was found that, in some settings, glacial erosion is insufficient to remove sufficient material to erase memory of pre-exposure (Fabel *et al.* 2002, Marquette *et al.* 2004, Briner *et al.* 2005, Phillips *et al.* 2006, Li *et al.* 2008). However, glacially over-ridden bedrock surfaces can provide important constraints on the thermal state of glaciers (Harbor *et al.* 2006). Cold-based ice, i.e. where ice is frozen to the ground and movement is accommodated by deformation within the ice, is non-erosive and can protect old landscapes from erosion (Fabel *et al.* 2002, Marquette *et al.* 2004, Briner *et al.* 2005, Phillips *et al.* 2006, Li *et al.* 2008). Wet-based ice, where glacial movement is taking place on the ice–bedrock interface, is erosive and can erase the cosmogenic memory of eroded bedrock surfaces. Therefore, bedrock surfaces are now increasingly being targeted to investigate the thermal regime of ice sheets, which can provide information on ice-sheet dynamics and its climatic forcing (Harbor *et al.* 2006, Jamieson *et al.* 2008, Li *et al.* 2008). Erratics found on glacially over-ridden bedrock usually provide reliable deglaciation ages (Fabel *et al.* 2002, Briner *et al.* 2006).

A note of caution on the preservation of glacial striations: anecdotal evidence suggests that glacial striations often best survive weathering under a soil/peat cover (John Gosse, personal communication). Thus, in areas where soil/peat cover is possible, this feature merits consideration concerning corresponding shielding.

Fluvially cut surfaces

River-cut surfaces may be abandoned by a river's continuing incision. If abandoned, fluvially smoothed/polished surfaces are preserved, and the time since their abandonment, as well as the time-integrated river incision rate can, in principle, be determined by exposure dating (Burbank *et al.* 1996, Ruszkiczay-Rüdiger *et al.* 2005, Schaller *et al.* 2005). In cases where these river-cut surfaces are found in a deeply incised gorge or in a locally complicated shielding situation (e.g. in a pot hole), accurate shielding corrections are imperative (Section 4.1.2).

To judge whether suspected river-cut surfaces were indeed carved by the river, rather than smoothed by weathering, it is beneficial to investigate modern river-cut features on the same lithology at the present-day river level (high-water level).

Wave-cut surfaces

Marine and lacustrine wave action can cut bedrock platforms that may be subsequently abandoned by uplift, or sea- or lake-level changes. These can be dated using cosmogenic nuclides (Stone *et al.* 1996b, Lifton *et al.* 2001, Alvarez-Marron *et al.* 2008). Potential issues that should be considered when sampling wave-cut platforms are whether a platform was sufficiently shielded from cosmic rays prior to excavation (particularly muons; Section 1.4), and to reliably identify rock surfaces that were smoothed by wave action rather than weathering. The former may be investigated via conceptual morphological reconstruction of the pre-existing landforms combined with a suitable sampling strategy (Fig. 4.14), the latter by examining and comparing with weathering features of rocks from the same lithology that were not affected by wave action.

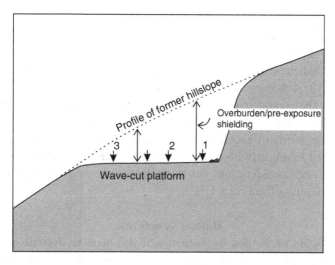

Fig. 4.14. The pre-erosional shielding of a wave-cut platform (i.e. incomplete shielding will result in nuclide inheritance) depends on the position of the sampling sites on the platform, and the difference in duration between pre-erosion and post-erosion exposure. The thickness of the shielding can be estimated by interpolating hill-slopes that pre-date the wave action. Samples near the cliff (1) usually have more pre-erosional shielding than those further away from it (3). Samples near the cliff, however, may be affected by talus of the cliff-face and may require a significant topographic shielding correction (f_{topo} is ~0.8, ~0.9 and ~0.98 for sites 1, 2, and 3, respectively; $m = 2.3$).

Wind-polished bedrock

Aeolian activity can remove significant overburden and expose erosion-resistant lithologies (Fodor *et al.* 2005, Youngson *et al.* 2005, Knight 2008). Mostly fine-grained lithologies resistant to weathering can preserve signs of aeolian activity (Lancaster *et al.* 2002, Viles and Bourke 2007). Aeolian activity may be discontinuous and linked to climatic change. Recurring arid or cold conditions can remove vegetation cover and episodically permit sand-drift and abrasion by sand. Thus, in principle, episodes of significant discontinuous aeolian erosion can be dated with exposure dating (Youngson *et al.* 2005). Pre-exposure, limited removal of overburden, and the preservation of original aeolian features through repeated aeolian cycles, are potential issues that should be considered when sampling aeolian landforms for exposure dating. The distinct nature of aeolian features on rocks (Viles and Bourke 2007) (see also Fig. 4.9) should allow the reliable exclusion/quantification of non-aeolian erosion on potential sampling sites.

Rockfall scar

Rockfalls may expose deeply shielded bedrock surfaces. The actual shielding depth prior to excavation is, however, usually unknown and unaccounted pre-exposure can affect calculated exposure ages. Depending on the lithology, rockfall scarps may or may not have unique markings that could indicate the preservation of an original uneroded scar surface. Due to the difficulties identifying uneroded scar surfaces, usually the debris of rockfalls is targeted for exposure dating, which is also easier to sample (Kubik *et al.* 1998, Hermanns *et al.* 2001). Sampling strategies and considerations for these deposits are not unlike those for moraine deposits (Section 4.1.3).

Human structures

Bedrock surfaces shaped by human activity, such as ancient quarries, can, in principle, be dated using cosmogenic nuclides. Methodological considerations would be similar to those for wave-cut surfaces (see above; Fig. 4.14). Further, tool marks on rock surfaces may be used to assess postexcavation erosion.

Fault scarps

Vertical tectonic movements can expose bedrock surfaces along fault scarps, which can be used to determine earthquake recurrence rates and/or slip rates, utilizing exposure dating (Mitchell *et al.* 2001, Benedetti *et al.* 2002, Benedetti *et al.* 2003, Palumbo *et al.* 2004). Bedrock normal fault scarps in carbonate rocks commonly preserve features indicative of fault movement; i.e. postmovement erosion can often be quantified or excluded (Mitchell *et al.* 2001, Benedetti *et al.* 2002, Benedetti *et al.* 2003, Palumbo *et al.* 2004).

Because co-seismic fault displacement is usually smaller than several times the attenuation path length of cosmic rays (Section 1.4), a significant pre-exposure component is essentially always present on bedrock scarps of normal faults. Exposure ages of fault scarps must therefore include a correction for pre-exposure during the time when the now-exposed footwall surface was still buried under the hanging wall (Mitchell *et al.* 2001, Benedetti *et al.* 2002, Benedetti *et al.* 2003, Palumbo *et al.* 2004). In practice, synthetic depth–concentration profiles predicted for fault movements (Fig. 4.15) are matched to the measured profiles from cosmogenic samples. The goodness of fit, considering the number of parameters included in the model to achieve a good fit, can constrain the most likely displacement histories (Mitchell *et al.* 2001, Palumbo *et al.* 2004).

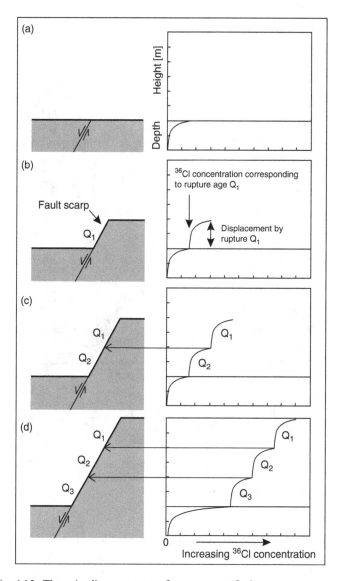

Fig. 4.15. The episodic movement of an emergent fault scarp on a normal fault can result in a resolvable fine-structure along a height–concentration profile up the scarp. The depicted scenario implies individual earthquake ruptures (Q1, Q2, Q3) in the order of 4 m, and a constant earthquake recurrence interval. In (d) the fault movement has ceased for three recurrence intervals. In quiescent periods a characteristic sub-surface concentration depth profile develops, the shape of which survives upon exposure on the fault scarp (subsequent exposure adds cosmogenic nuclides to all locations on the exposed scarp at an identical rate). After Palumbo *et al.* (2004).

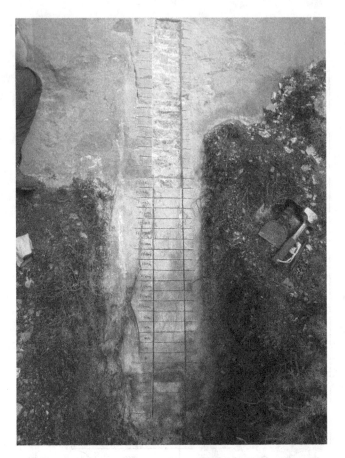

Fig. 4.16. Sampling of a profile along a limestone fault scarp, Abruzzo, Italy. The depicted section covers the transition from the exposed to the soil covered part of the fault scarp.

The shape of the depth–concentration profiles of individual events (Fig. 4.15) is dependent on the nature of the hanging wall. Often the base of the fault scarp is covered in soil, colluvium or tectonic breccia (Fig. 4.16), which are less dense than the intact bedrock of the scarp lithology, affecting the attenuation path length for cosmogenic-nuclide production. For accurate modelling results, the actual density of the hanging wall should be investigated and taken into account.

Tectonic movements can displace moraines (Matmon *et al.* 2006, Meriaux *et al.* 2009), and inactivate fluvial terraces and alluvial fans, which can be dated by exposure dating (Siame *et al.* 1997, Hetzel *et al.* 2002b,

Daeron *et al.* 2004, González *et al.* 2006, Ryerson *et al.* 2006, Frankel *et al.* 2007, Carizzo *et al.* 2008). The age determination of these features allows constraining of rates of lateral and/or vertical fault movements. Sampling strategy and methodology are the same as would be used to assess the age of formation or abandonment of these depositional features (Section 4.1.3), and is therefore not repeated here.

4.2 Burial dating

In contrast to exposure dating (Section 4.1), which relies on the continuous accumulation of cosmogenic nuclides at the surface, burial dating relies on the differential decay of cosmogenic nuclides; in situations where previously exposed samples become temporarily or permanently shielded from cosmic rays. Burial conditions can occur when sediments are washed into caves or are buried by subsequent deposition, or when rocks are temporarily or permanently covered by water, ice, till, dunes or volcanic material (Granger 2006). An advantage of burial dating over exposure dating is that erosion of the buried sediments or surfaces is usually not an issue, as they are protected from the weathering and eroding agents that are active at the surface.

Burial dating exploits the different half-lives of cosmogenic radionuclides in that the initial (preburial) ratio of two radionuclides $R_{AB}(0)$ (Eqn. (4.13)) changes according to the corresponding half-lives (λ_A, λ_B) with increasing burial duration t_b (Klein *et al.* 1986, Lal 1991, Granger and Muzikar 2001, Balco and Rovey 2008):

$$R_{AB}(t_b) = \frac{C_A(t_b)}{C_B(t_b)} = R_{AB}(0) \; e^{-t_b(\lambda_A - \lambda_B)} \qquad (4.22)$$

The burial duration t_b can thus be calculated as:

$$t_b = \frac{-\ln\left(\frac{R_{AB}(t_b)}{R_{AB}(0)}\right)}{(\lambda_A - \lambda_B)} \qquad (4.23)$$

Concentrations $C_A(t_b)$ and $C_B(t_b)$ and their ratio $R_{AB}(t_b)$ denote measured values. An essential prerequisite of this method is that the initial ratio $R_{AB}(0)$ is known. Equation (4.22) is valid for situations where postburial production at depth is negligible.

The pair ^{26}Al and ^{10}Be is particularly well suited for burial dating. This is because: (i) they are produced in the same mineral (quartz), (ii) their production ratio is largely independent from altitude and latitude, and

(iii) does not vary substantially with depth in a rock (Brown et al. 1992). Features (ii) and (iii) have, as a consequence, an initial, preburial ratio that is usually well constrained and close to the spallogenic production ratio at the surface (Granger and Muzikar 2001, Granger 2006, Balco and Rovey 2008). The latter assumption (iii) requires consideration/ confirmation for each application (Granger and Muzikar 2001, Granger 2006, Balco and Rovey 2008). In landscapes where the buried material is supplied by erosion at rates faster than $10 \, \mathrm{m \, Ma^{-1}}$, the initial $^{26}\mathrm{Al}/^{10}\mathrm{Be}$ ratio is relatively insensitive to the erosion rate (Balco and Rovey 2008).

In line with the current application of burial dating (Granger 2006), mainly the pair $^{26}\mathrm{Al}$ and $^{10}\mathrm{Be}$ is considered in this section. Other potential nuclide pairs may have depth-dependent production ratios (e.g. $^{36}\mathrm{Cl}/^{10}\mathrm{Be}$) due to significant non-spallogenic production pathways (e.g. via thermal-neutron capture; Section 1.5). Consequently they may have variable and unconstrained initial ratios $R_{AB}(0)$ in buried material, rendering them useless for burial dating. Throughout Section 4.2, subscripts A and B in the equations can be read as $^{26}\mathrm{Al}$ and $^{10}\mathrm{Be}$, respectively.

The useful time-range of burial dating is determined by the half-life of $^{26}\mathrm{Al}$ and the analytical precision of its measurement. Assuming an experimental uncertainty for the determination of $R_{AB}(t_b)$ (Eqns. (4.22) and (4.23)) of 5%, the uncertainty in burial ages is about 100 ka; thus the lower limit of the usefulness of the method is on the order of 100 ka (Granger and Muzikar 2001). The upper limit is on the order of 5 Ma, and caused by the depletion of $^{26}\mathrm{Al}$ below analytical detection levels. Depending on the amount of the originally buried $^{26}\mathrm{Al}$, this time-range may be shorter in practice (Granger and Muzikar 2001).

Because burial dating relies on the decay of $^{26}\mathrm{Al}$ and $^{10}\mathrm{Be}$, the accurate knowledge of their decay constants and corresponding half-lives (Section 2.1) is important. However, different values of the half-life of $^{10}\mathrm{Be}$ are currently in use (Section 2.3.1). The revised half-life of $^{10}\mathrm{Be}$ is 1.36 ± 0.07 Ma (Nishiizumi et al. 2007) is not yet universally adopted, and the previously widely used value of 1.51 ± 0.06 Ma (Hofmann et al. 1987) is still often used. At the time of writing, two independent evaluations of the $^{10}\mathrm{Be}$ half-life were concluded, both reporting a value of 1.39 Ma, with an stated uncertainty of about 1% (Chemeleff et al. 2009, Korschinek et al. 2009). These new values are in agreement with the revised value of Nishiizumi et al. (2007). Irrespective of which half-life is actually used for burial dating, the choice must be clearly stated, and concentrations as well as production ratios of $^{26}\mathrm{Al}/^{10}\mathrm{Be}$, which depend on the assumed half-life, must be used consistently (Granger 2006, Nishiizumi et al. 2007,

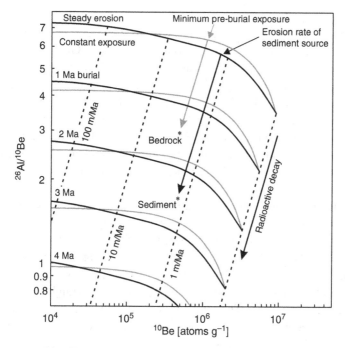

Fig. 4.17. ^{26}Al/^{10}Be vs. ^{10}Be diagram for situations including burial of samples (building on principles explained for Fig. 4.3). Radioactive decay after burial of samples changes their cosmogenic nuclide inventory on trajectories similar/parallel to those indicated. Buried sediments will develop starting from the steady state erosion line, at a point that is indicative for the erosion rate of the sediment source. Bedrock samples may develop from the zero-erosion line at a point commensurate to their prior exposure. The sample trajectories show the first burial event experienced by a hypothetical sample. Particularly for bedrock samples, repeated burials and exposures are possible, resulting in more complex trajectories below the island of steady state erosion (e.g. Fabel *et al.* 2002). After Granger and Muzikar (2001).

Balco and Rovey 2008, Balco *et al.* 2008). In this book the revised half-life of Nishiizumi *et al.* (2007) is used, with a corresponding ^{26}Al/^{10}Be production ratio of 6.75 (Balco and Rovey 2008).

4.2.1 Fast and complete burial

For samples that are quickly buried and subsequently fully shielded from cosmic rays, the ^{26}Al/^{10}Be vs. ^{10}Be diagram introduced earlier (Section 4.1; Fig. 4.2) can be usefully expanded to illustrate burial histories (Fig. 4.17), and corresponding burial ages t_b can be calculated using Eqn. (4.23).

Fast, complete and permanent burial is realized in cases where sediments are flushed into deep cave systems, deposited under thick sediment sequences or in deep water (Granger *et al.* 1997, Granger and Smith 1998, Granger *et al.* 2001, Haeuselmann *et al.* 2007, Kong *et al.* 2009).

Depending on the nuclide concentration present in the sediments at burial and the subsequent burial duration, shielding should be in excess of 30 m (rock equivalent mass) to maintain that a site is effectively shielded from cosmic rays (Granger and Muzikar 2001). The condition of fast burial is met in all cases where the time-span over which shielding increases to complete shielding is much shorter than the subsequent burial duration.

Prominent examples for exposure dating of cave sediments are studies that trace the deepening of cave systems in response to river incision (Granger *et al.* 1997, Granger and Smith 1998, Granger *et al.* 2001, Stock *et al.* 2004, Haeuselmann *et al.* 2007), and dating of cave sediments associated with hominid finds (Partridge *et al.* 2003, Gao *et al.* 2009). The sampling of quartz-bearing cave sediments is straightforward (i.e. bag and mark them). The only important consideration is to avoid situations where samples could be contaminated by material from older or younger deposition events.

A special case of complete shielding is where repeated advances of cold-based ice sheets (Section 4.1.4) over bedrock surfaces, which can cause repeated full shielding, are interrupted by periods of full exposure. In these cases of complex burial histories, evolution trajectories, such as depicted on Fig. 4.17, provide minimum ages of such surfaces (Fabel *et al.* 2002, Marquette *et al.* 2004, Phillips *et al.* 2006, Li *et al.* 2008).

4.2.2 Slow and/or incomplete burial with variable inheritance

In situations where burial is relatively shallow (<30 m), as is commonly the case with buried sediments and soils, postburial production by deeply penetrating muons remains significant, and needs to be considered (Granger and Muzikar 2001, Balco and Rovey 2008). Furthermore, soils may have a non-trivial exposure history prior to burial. There are two sampling strategies to tackle these situations: depth profiles in soils or sediments (Granger and Smith 2000, Wolokowinsky and Granger 2004, Balco *et al.* 2005, Granger 2006, Balco and Rovey 2008), or sampling multiple clasts at one depth level in the sediment (Balco and Rovey 2008).

Depth profiles for burial dating

Burial dating of palaeosoils developed on sediment surfaces requires a refined approach. The presence of palaeosoils indicates that a buried sediment surface was exposed for a significant period prior to burial. Thus, the cosmogenic nuclide inventory just prior to burial is determined by the locally produced $R_{AB}(0)$ *and* any inherited concentrations $C_{A,inh}$ and $C_{B,inh}$ from the previous exposure/burial history of the sediment (Eqn. (4.10)). If the sediment has a protracted history of partial/ temporary burial and radioactive decay during transport and remobilization prior to the latest deposition, it will start off with a ^{26}Al depletion relative to what would be expected from the surface production ratio $R_{AB}(0)$ and the ^{10}Be concentration (i.e. with an inherited burial age) (Granger 2006, Balco and Rovey 2008). Thus, if this inherited component is not considered, calculated burial ages will overestimate the actual burial age (Balco and Rovey 2008). A novel approach, using an isochron method, can resolve this ambiguity and produce reliable burial ages, if the inherited component is homogenous throughout the sampled profile (Balco and Rovey 2008). The main features of this approach are described below, while its derivation is described in detail by Balco and Rovey (2008).

Independent of sample depth, the ^{26}Al concentration C_A in a buried palaeosoil is described by:

$$C_A(t_b) = R_{AB}(0) \; e^{-t_b(\lambda_A - \lambda_B)} C_B(t_b) + [C_{A,inh} e^{-t_b\lambda_A} - R_{AB}(0) \; C_{B,inh} e^{-t_b\lambda_A}]$$

$$(4.24)$$

(Balco and Rovey 2008). The term in square brackets describes the inheritance; if inheritance is zero, Eqn. (4.24) is equivalent to Eqn. (4.22). Equation (4.24) is a linear relation between ^{10}Be and ^{26}Al concentration, valid for all depths in a profile (Balco and Rovey 2008). Thus, the results of cosmogenic nuclide measurements of samples from various depths in a palaeosoil profile will lie on a line with a slope of $R_{AB}(t_b)$ $e^{-t_b(\lambda_A - \lambda_B)}$ (i.e. Eqn. (4.22)) passing through the point ($C_{A,inh} e^{-t_b\lambda_A}$, $C_{B,inh} e^{-t_b\lambda_B}$). The measured slope $R_{AB}(t_b)$ of such a data array depends on the unknown burial duration t_b, the known decay constants and surface production ratio. The burial age is then determined by Eqn. (4.23).

For cases with no significant inheritance, the measured nuclide ratio $R_{AB}(t_b)$ in each sample can be used to calculate the burial age, and data arrays should lie on lines that go through the origin in a linear ^{26}Al vs. ^{10}Be diagram (Fig. 4.18). Data arrays with inheritance are rotated and do

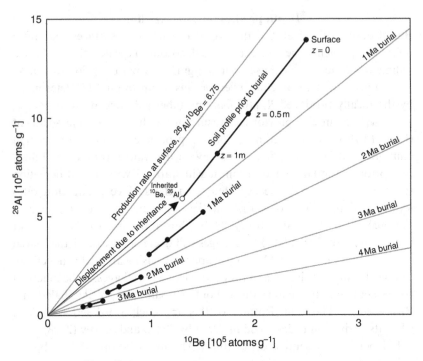

Fig. 4.18. ^{26}Al–^{10}Be isochron diagram after Balco and Rovey (2008). Samples with a simple burial history, i.e. without inheritance and/or postburial production, will rotate along trends indicated by the grey lines. Sets of samples with a different but related pre burial production and identical average pre-exposure (for instance samples of a buried paleo-soil profile) will evolve on displaced lines, parallel to the simple buried samples. In both cases the slopes of the trends defined by the sets of samples are used to calculate burial ages (eq. 4.22). Sets of samples with variable pre-exposure and identical post burial production (e.g. sediment samples from the same depth) behave in a similar fashion and can be dated accordingly (see text for discussion).

not pass through the origin of such a diagram – thus the presence or absence of significant inheritance can be identified and quantified. The assumption of a homogenous inheritance throughout the depth profile is an essential prerequisite, which is tested by the linear fit of the data array. Samples that violate the assumption will not lie on a line (Balco and Rovey 2008). For practical use, it is important that samples have a large enough spread in concentrations, such that the slope of the data array can be determined accurately. This is achieved by sampling profile lengths in the order of 1.5 to 2 m (Balco and Rovey 2008).

Naturally, this isochron method is also applicable when there is no palaeosoil horizon indicative for a non-trivial exposure history. It will pick up inheritance if it is significant, and if it is not, it will provide the same result as using individual samples' nuclide ratios (Section 4.2.1; Balco and Rovey 2008).

Alternative approaches for burial dating of sediments with a simple two-stage exposure–burial history (i.e. exposure during erosion, brief transit in the sediment routing system, burial on deposition) involve the parameter fitting of models to measured concentration profiles in long depth profiles (5–10 m), in order to produce models that best describe the observed profile (Granger and Smith 2000, Wolokowinsky and Granger 2004).

In all cases, where depth profiles in buried sediments or palaeosoils are sampled for burial dating, postburial nuclide production needs to be corrected for using the measured or inferred burial depth, the relative difference of shielding depth between samples and, if applicable, the temporal evolution of the thickness of the overburden. The relevant methods and concepts of postburial production correction, which are often tailored for specific situations, are described in detail in Granger and Muzikar (2001), Granger (2006) and Balco and Rovey (2008).

Isolevel sampling of buried sediments
Sediment units in fluvial terraces are usually not sufficiently thick to completely shield even their lowest horizons from cosmic rays. Postburial production is therefore important in most cases. The isochron method introduced above can be used to tackle the issue of postburial production in sediments with a simple two-stage exposure history (Balco and Rovey 2008). For these sediments, the surface production ratio $R_{AB}(0)$ in the source region defines the initial buried ratio (Granger and Muzikar 2001, Granger 2006, Balco and Rovey 2008), which is therefore well constrained, unlike in the previously discussed case for palaeosoils. Further, sediment clasts sampled at the same depth level below the surface will all have the same nuclide concentration from postburial production ($C_{A,pb}$ and $C_{B,pb}$). In any sample from such a sediment the ^{26}Al concentration is defined by:

$$C_A(t_b) = R_{AB}(0)\, e^{-t_b(\lambda_A - \lambda_B)} C_B(t_b) + [C_{A,pb} e^{-t_b\lambda_A} - R_{AB}(0)\, C_{B,pb} e^{-t_b\lambda_A}]$$

$$(4.25)$$

(Balco and Rovey 2008), which is very similar to Eqn. (4.24). Thus, analogous to the discussion of Eqn. (4.24), the results of cosmogenic nuclide measurements of a set of samples from the same depth below the surface will lie on a line with a slope of $R_{AB}(t_b)e^{-t_b(\lambda_A-\lambda_B)}$ passing through the point $(C_{A,pb}e^{-t_b\lambda_A}, C_{B,pb}e^{-t_b\lambda_B})$. The measured slope $R_{AB}(t_b)$ of such a data array depends on the sought burial duration t_b, and the known decay constants and surface production ratio. The burial age is then determined by Eqn. (4.23). The result is independent of the postburial exposure history, i.e. independent of the time–depth history of the sampled sediment layer after its original deposition (Balco and Rovey 2008).

In cases without significant postburial production, the measured nuclide ratio $R_{AB}(t_b)$ in each sample could be used to calculate the burial age, and data arrays lie on lines that go through the origin in a linear ^{26}Al vs. ^{10}Be diagram (Kong *et al.* 2009) (Fig. 4.18). Data arrays with significant postburial production are rotated and do not pass through the origin of such a diagram (Balco and Rovey 2008). Thus the presence or absence of significant postburial production can be identified and quantified (though the latter is not necessary to obtain t_b, see above).

The assumption of a simple two-stage exposure history of the sediment samples is an essential prerequisite for the application of the isochron method, which is tested by the linear fit of the data array. Samples that violate the assumption will not lie on a line. The necessary spread of concentrations to obtain accurate estimates of the slope of the data array is commonly realized in coarse sediment clasts (Brown *et al.* 1995, Repka *et al.* 1997, Balco and Rovey 2008, Codilean *et al.* 2008). Clasts need to be large enough to permit single clast analysis (see Section 2.4.1). The isochron method of Balco and Rovey (2008) has the potential to allow the age determination of previously undatable continental clastic sediments (Fig. 4.19).

4.2.3 Complex burial histories

The methods described in the previous two sections usually aim at constraining the age of discrete deposition/burial events. Most surface materials may have complicated exposure histories. Individual surface grains may have been repeatedly exposed, and/or partially or fully shielded from cosmic rays. While in these cases it is not possible to derive a distinct age for the processes operating, it is usually possible to constrain their minimum duration. The interpretation of complex burial histories follows the conceptual approach shown in Fig. 4.17 for repeatedly exposed surfaces.

Fig. 4.19. With the new isochron method of Balco and Rovey (2008) erosive remnants of old (up to Pliocene) continental sediments become viable targets for dating with cosmogenic nuclides. In the depicted example (former valley fill sediments in the Ugab catchment, Namibia) samples from the top of the caves at the base of the sediment pile could be a suitable target for this new form of burial dating (work in progress).

Both the fate of grains in (self-) shielding material (e.g. dunes), and shielded material (e.g. gravel on a desert plain over-ridden by dunes) can be investigated. In the case of moving (self-) shielding material, the minimum cumulative transport time can be inferred, and in the case of the shielded material its minimum residence time at the surface. Examples include transport of sand along coastal shorelines (Nishiizumi *et al.* 1993, Boaretto *et al.* 2000), the protracted evolution of desert pediments (Bierman and Caffee 2001), the exposure history of individual clasts on desert pavements (Klein *et al.* 1986) and the detection of transient soil cover (Albrecht *et al.* 1993, Braucher *et al.* 2000). Also the aforementioned multiple over-riding by cold-based ice sheets of glacial bedrock belongs to the family of complex exposure histories (Fabel *et al.* 2002, Marquette *et al.* 2004, Miller *et al.* 2006, Phillips *et al.* 2006, Anderson *et al.* 2008, Li *et al.* 2008).

4.3 Erosion/denudation rates

Erosion of a landscape continuously brings new material to the surface. Mineral grains in transit from a shielded position in the subsurface to the fully exposed position at the surface will accumulate a cosmogenic nuclide inventory proportional to the transit time, i.e. proportional to the erosion rate.

In Section 4.1.1 it was shown that the cosmogenic nuclide concentration in a mineral grain brought to the surface ($z = 0$) at a steady state erosion rate ε (Lal 1991) is described by:

$$C_{\text{total}}(z) = \sum_i \frac{P_i(z)}{\lambda + \rho\varepsilon/\Lambda_i} \tag{4.12}$$

for radionuclides, and thus by

$$C_{\text{total}}(z) = \sum_i \frac{P_i(z)}{\rho\varepsilon/\Lambda_i} \tag{4.26}$$

for stable nuclides, or situations where radioactive decay is negligible. In the context of cosmogenic erosion-rate determinations the term Λ/ρ is often replaced by z^*, the mean attenuation path length (Section 1.4), which is ~ 60 cm in silicate rocks (Lal 1991, von Blanckenburg 2005).

For surfaces that erode sufficiently slowly such that radionuclides produced at great depth by muons have decayed before they reach the surface (i.e. erosion rates <10 m Ma^{-1} for ^{10}Be), Eqn. (4.12) can be simplified to

$$C(0) = \frac{P(0)}{\lambda + \rho\varepsilon/\Lambda} \tag{4.27}$$

providing a practical estimate for spallogenic nuclide concentrations for most eroding surfaces. The erosion rate ε is then given by

$$\varepsilon = \left(\frac{P(0)}{C(0)} - \lambda\right)\frac{\Lambda}{\rho} = \left(\frac{P(0)}{C(0)} - \lambda\right)z^* \tag{4.28}$$

Thus, if we measure the concentration $C(0)$ of a spallogenic cosmogenic nuclide (decay constant λ; $\lambda = 0$ for stable nuclides) at the surface, and we know its production rate at the surface $P(0)$, the density of the eroding material ρ and the attenuation coefficient Λ for the nucleonic component of the cosmic radiation ($\Lambda \sim 160$ g cm^{-2}; see Section 1.4), we can calculate the steady-state erosion rate ε. Situations with temporally changing erosion rates are discussed in Section 4.3.1.

Erosion vs. denudation

While it is erosion, i.e. the physical removal of material from the Earth's surface, which usually brings mineral grains to the surface, cosmogenic nuclides are actually sensitive to the denudation rate of the material above them, i.e. the rate of mass loss. The denudation is the sum of mass loss due to physical erosion and chemical weathering (dissolution).

The relative importance of chemical weathering can be assessed by enrichment of insoluble elements in a soil. Zirconium is customarily used as a proxy for weathering intensity. In acidic silicate rocks, Zr is mainly hosted in the mineral zircon ($ZrSiO_4$), which is extremely resistant to weathering, and its concentration will increase with weathering intensity relative to less resistant minerals. By comparing the Zr content of the soil to that of the unweathered bedrock, the relative enrichment factor f_e can be determined; $f_e = [Zr_{soil}]/[Zr_{bedrock}]$ (Granger and Riebe 2007). The erosion rate ε and denudation rate D are related by

$$\varepsilon = \frac{D}{\rho \cdot f_e} \qquad (4.29)$$

Enrichment factors for soils on granitic rocks may range between 1.2 and 2.5 (Granger and Riebe 2007), with high values in tropical areas.

For cosmogenic applications, the denudation rate and erosion rate are essentially equivalent in many instances. However, in situations with intense chemical weathering (e.g. in tropical areas), mass loss can also occur deep in the subsurface, and the cosmogenic nuclide abundance in grains at or near the surface will not reflect weathering mass losses occurring at depths that are significantly greater than the mean attenuation path length z^*. In these cases, the denudation rate of the landscape is larger than the denudation rate recorded by the nuclide concentrations at the surface. Thus, while cosmogenic nuclide abundances always relate to the erosion rate of a landform (i.e. surface-lowering rate of material with a known density) they do not always record its denudation (mass loss) if it occurs at significant depth z ($z > z^*$; Fig. 4.20). In many instances, however, the terms erosion and denudation remain interchangeable if effects of weathering occur at shallow depth only ($z < z^*$), and are taken into account (Eqn. (4.29)). Keeping these limitations in mind; the term denudation rate is mainly used in the context of basin-scale rates, which are discussed in the following. However, to maintain readability, and to avoid unnecessarily repeating equations, erosion ε is continued to be used in equations throughout.

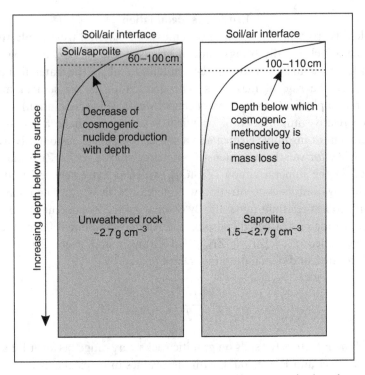

Fig. 4.20. Chemical weathering can remove significant mass in the subsurface via dissolution. In tropical areas weathered saprolite can extend for tens of metres below the ground. Cosmogenic nuclides are insensitive to mass loss (denudation) that occurs below ~60–100 cm (depending on density of soil/rock). Consequently cosmogenic nuclide concentrations near the surface do not reliably reflect the area's denudation rate. Cosmogenic nuclides will, however, record surface lowering of the soil/air interface (erosion) reliably, even in situations with significant weathering.

4.3.1 Basin-scale denudation rates

An important feature of cosmogenic-nuclide-derived erosion rates is that the above equations (Eqns. (4.12) and (4.26)–(4.27)) can be up-scaled, i.e. used for sediments leaving a catchment and calculating catchment-wide erosion rates $\bar{\varepsilon}$ (Brown *et al.* 1995, Bierman and Steig 1996, Granger *et al.* 1996) (Fig. 4.21); the average nuclide concentration in the sediment $\bar{C}$ is then replacing $C(0)$ and the average nuclide production rate $\bar{P}$ in the catchment replacing $P(0)$ (Brown *et al.* 1995, Bierman and Steig 1996, Granger *et al.* 1996):

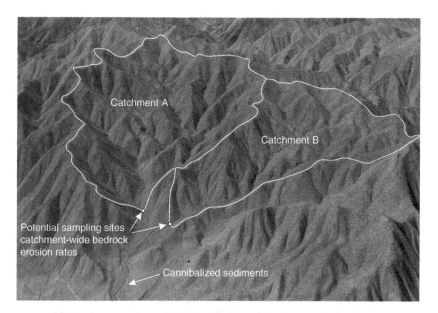

Fig. 4.21. Cosmogenic nuclide analyses of sediment samples from the active channels leaving catchments A and B can yield their corresponding average denudation rates. Below the indicated sampling sites the channels incise into their own sediments, partially remobilizing them (cannibalizing); such situations should be avoided for sampling (see text).

$$\overline{C} = \frac{\overline{P}}{\lambda + \overline{\rho}\,\overline{\varepsilon}/\Lambda} \qquad (4.30)$$

The nuclide concentration of sediment is measured as for any other sample (Sections 2.4 and 2.5), while the derivation of the average production rate needs to account for the basin's hypsometry, which is discussed later in this section.

The application of these equations to basin-wide erosion rates works accurately only if the following six conditions are met:

(1) Erosion in the catchment is constant over the period over which the cosmogenic nuclides average denudation. The averaging time T_{ave} is the time required to remove one mean attenuation path length z^*; i.e. $T_{ave} = z^*/\varepsilon$ (Lal 1991), after which most of the cosmogenic 'memory' of previous conditions is effectively erased (von Blanckenburg 2005). This condition may be violated in areas where landslides or other mass-wasting processes contribute significantly to an area's erosion

rate. In such cases, samples may be biased by material exposed from depth by a recent proximal landslide. However, by choosing a sufficiently large area, to ensure that the stochastic nature of landslide occurrence is reflected in an average sediment sample, this condition can be met, even in areas with significant landslide activity (Niemi *et al.* 2005, von Blanckenburg 2005). The resilience of the method to temporally variable erosion rates is discussed later in this section.

(2) The target mineral in the sediment must represent all lithologies in the catchment; in other words, all lithologies in the catchment must contribute to the sediment load of the target mineral commensurate to their particular erosion rates. For example, if a catchment contains a mixture of limestone and sandstone, and quartz is the target mineral, only information for the sandstone erosion rate is obtained. Information on the limestone erosion rate may be inferred by morphological comparison to the sandstone, but not directly from the sediment analysis.

A more subtle variation of this condition is that quartz contents may vary in quartz-bearing lithologies, e.g. granites may contain between 20 and 60% of quartz (Streckeisen 1973). Quartz-rich portions of a nominal granitoid catchment can contribute more to the quartz sediment budget, and bias the erosion rate accordingly if the quartz-rich lithologies have a different erosion rate compared to the quartz-poor lithologies. The accuracy of the denudation rate obtained is thus limited by the variability of the target-mineral abundance in the catchment.

(3) The target mineral must have the same grain size in all lithologies of the catchment. Analogous to the example above, if a catchment contains conglomerates and sandstones, and pebbles are collected as sediment samples, only information on the erosion of the conglomerate is obtained. If there is a contrast in erosion resistance or weathering behaviour, the results will be biased accordingly (Brown *et al.* 1995, von Blanckenburg 2005). The sampling strategy should be designed to ensure that representative minerals and grain sizes are extracted from the sediments, and/or to identify anomalous lithologies (in terms of erosion behaviour and petrography) that may contribute to the sample.

(4) Mass loss from the basin should occur primarily by surface lowering (Bierman and Steig 1996), i.e. not by chemical weathering in the (deep) subsurface. If weathering is deep and significant, cosmogenic nuclides will not reflect the area's denudation rates, as they are not

sensitive to mass-loss occurring below the absorption length $z*$ for cosmic radiation (Fig. 4.20).

In vertically mixed soils, quartz may be enriched by selective dissolution (Small *et al.* 1999, Granger and Riebe 2007). If not accounted for, calculated denudation rates will be too low (Small *et al.* 1999, Riebe *et al.* 2001, 2004). Correction procedures (see below) utilize the immobility of zirconium to assess the relative intensity of chemical weathering (Riebe *et al.* 2001, 2004).

(5) Transit times of sediments through the catchment should be short when compared to the erosional timescale (see 6). If violated, sediments reflecting a different erosional regime may be sampled rather than those representing the current regime. Also radioactive decay during transit may be significant (see 6). This condition should be investigated particularly in situations where sediment storage is indicated by the presence of thick colluvium and/or sedimentary terraces, and in cases where the drainage system is running over its own sediments.

(6) The time required to remove one absorption length $z*$ should be short compared to the half-life of the cosmogenic nuclide applied, i.e. $z*/\varepsilon << T_{1/2}$. If this condition is violated in catchments with a spatially variable erosion rate, areas with lower erosion rates will be relatively under-represented in a sample (Bierman and Steig 1996). In practice, this limits the use of [10]Be to catchments eroding faster than 0.3 m Ma^{-1} (von Blanckenburg 2005). However, this is a condition that is met in many geological settings. For ^{14}C, at the low end of the half-life spectrum of cosmogenic nuclides, the corresponding limit is 80 m Ma^{-1} (von Blanckenburg 2005), which is rarely achieved. Thus, the application of ^{14}C for catchment-wide denudation rates is problematic and requires the investigation of spatial homogeneity of denudation rates (if they are homogenous ^{14}C can be applied, even in settings where ε <80 m Ma^{-1}).

In many natural situations one or more of the above conditions are not fully met. Sometimes this may mean the method cannot be applied, but often it simply limits the accuracy of the calculated denudation rates, without necessarily destroying its applicability. It is important to bear in mind that long-term denudation rates are generally difficult to derive with alternative, non-cosmogenic methods, and that arguments on denudation rates are hinged commonly on the order of magnitude of rates, rather than on several tens of percent difference (von Blanckenburg 2005). The above conditions, their partial violation in nature (e.g. conditions 2 and 3), and issues concerning the determination of average production rates (see below) probably limit the general accuracy of the method to ±20–30%. If the application requires a better accuracy, all the above conditions must be strictly met, and production-rate parameters must be known accurately. The latter are discussed below.

Deriving catchment-wide production rates

In order to calculate catchment-wide denudation rates, the average catchment production rate of cosmogenic nuclides needs to be derived. Production rates of cosmogenic nuclides are a function of altitude and latitude (Chapters 1 and 3). Altitudinal variations may be significant in many catchments, whereas relative latitudinal variations are not significant in the context of production-rate variations for all but the largest catchments, and are usually not considered. To account for the range in altitude, area-weighted elevation bands or digital elevation models are usually used to compute average nuclide production in a catchment (von Blanckenburg 2005). A simpler approach, using the mean latitude and the mean altitude of a catchment, often yields very similar results in all but high-relief terrains (von Blanckenburg 2005).

Besides altitudinal variations, topographic shielding can affect the average nuclide production rate in catchments. In canyon-like settings, topographic shielding can reduce production rates by up to 15–20% near the bottom of a canyon (Codilean 2006); average catchment-wide shielding will be less. Generally, in areas where the landscape is defined by hill slopes below the threshold for detachment limited transport, i.e. having hill-slope angles $<30°$ (Binnie *et al.* 2007), the maximum topographic shielding at the bottom of the topography is $<15\%$ (Gosse and Phillips 2001), and much lower for the catchment average. In a landscape defined by 20° hill-slope angles, the maximum shielding is less than 3% (Gosse and Phillips 2001). Using DEM-based calculations, the spatially resolved and mean shielding factors of a catchment can be determined (Binnie *et al.* 2006, Codilean 2006, Norton and Vanacker 2009).

Averaging timescales and temporally variable erosion rates

As introduced earlier, the time over which cosmogenic nuclides average catchment-wide denudation rates is the time required to remove one absorption length z^*, i.e. $T_{ave} = z^*/\varepsilon$, with z^* being $\sim 60\,cm$ in rocks or $\sim 1\,m$ in soil, for nuclides produced predominantly by spallation

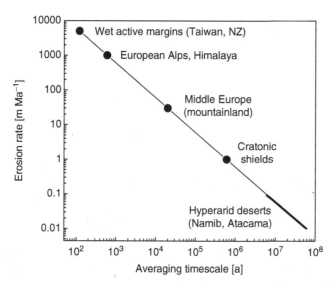

Fig. 4.22. The averaging timescales for cosmogenically derived erosion rates depend on the rates themselves (see text). Fast erosion goes hand in hand with short averaging times and vice versa. Examples are from various tectonic settings (von Blanckenburg 2005) and climatic zones (Dunai *et al.* 2005; van der Wateren and Dunai 2001). Format after von Blanckenburg (2005).

(von Blanckenburg 2005). Since they are dependent on the erosion rate, integration times can vary between 10^2 and 10^7 years for different tectonic and/or climatic settings (Fig. 4.22). Variability of erosion rate timescales both shorter or longer than the integration time is not resolved (Schaller and Ehlers 2006, Binnie *et al.* 2008). The timescale of validity of catchment-wide erosion rates can be extended beyond the method's inherent time limits using additional, external evidence (von Blanckenburg 2005).

The first condition given for obtaining accurate results for catchment-wide erosion rates was that of temporally constant erosion rates. In situations where this condition is violated, the cosmogenic nuclide concentration in the sediment will lag the actual erosion rate. To illustrate these effects, the most extreme situation, a step change, is considered in the following. The concentration of a cosmogenic nuclide in a surface that underwent a step change in erosion rate from ε_1 to ε_2 at time $t = 0$ is described by:

$$C(t) = \sum_i [C_{1,i} + (C_{2,i} - C_{1,i}) \cdot (1 - e^{-t(\varepsilon_2 \rho/\Lambda + \lambda)})] \qquad (4.31)$$

(Granger and Riebe 2007); with

$$C_{1,i} = \sum_i \frac{P_i(0)}{\lambda + \rho\varepsilon_1/\Lambda_i} \tag{4.32}$$

and

$$C_{2,i} = \sum_i \frac{P_i(0)}{\lambda + \rho\varepsilon_2/\Lambda_i} \tag{4.33}$$

(Granger and Riebe 2007). For nuclides dominated by spallogenic production, the equations simplify accordingly ($i = 1$; see Eqn. (4.27)). The response of the concentration $C(t)$ to changes in erosion rates is gradual, with the concentration asymptotically approaching the new steady-state equilibrium value. The time required to reach the new steady state depends either on the radioactive decay of the inherited concentration memory or the removal of one absorption length z^*, whichever is achieved fastest (Granger and Riebe 2007) (Fig. 4.23).

In the context of recent human disturbance, the considerable time lag between changes in erosion rates and when they actually show up in the cosmogenic nuclide record of sediments is usually a benefit. Memory of preanthropogenic erosion rates can be preserved, even in regions that suffered recent moderate erosion (i.e. <60 cm rock or <1 m soil). The cosmogenically derived erosion rates can be compared to current erosion rates obtained with traditional techniques (e.g. erosion plot studies, sediment traps), establishing the absence or presence of significant anthropogenic disturbance of a system (Hewawasam *et al.* 2003). Such studies also allow the establishment of the natural background erosion rate in regions affected by human activity (Vanacker *et al.* 2007).

Vertically mixed soils and regoliths

Most studies on catchment-wide erosion rates are in soil- and/or regolith-mantled landscapes. In particular, soils are commonly vertically mixed by bioturbation (e.g. tree throw, burrowing animals), which may affect the anticipated cosmogenic-nuclide concentration at the surface.

In the end-member situation where soils/regoliths are thoroughly mixed, but not removed by erosion, the cosmogenic-nuclide concentration in surface material simply reflects the mean concentration of a vertically mixed portion of the depth–concentration profile (Fig. 4.24). If such a vertically mixed soil profile starts to be eroded, material with this mean concentration will be removed, rather than material with the

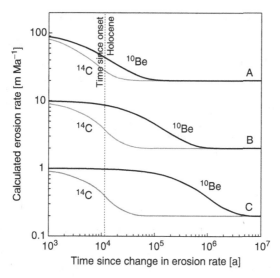

Fig. 4.23. After changes in erosion rates it may take a considerable time until the cosmogenic nuclide inventory of sediments regain steady state, and reflect the new erosion rate. The depicted examples are for a five-fold step change of erosion rates from 100, 10 and 1 m/Ma to 20, 2, and 0.2 m/Ma, respectively. The change occurred at t = 0; the diagram shows erosion rates that would be calculated from measured nuclide concentrations between times t = 1 ka and t = 10 Ma (they should show 20, 2, or 0.2 m/Ma, if there were no time-lag). In-situ produced ^{14}C responds faster to changes than ^{10}Be because of its short memory due to fast radioactive decay. Changes that may have occurred at the onset of the Holocene may be just traced with ^{14}C; whereas ^{10}Be will not come near the actual erosion rate in situations with slow to moderate erosion rates (see also Fig. 4.22 for typical rates). This limitation should be kept in mind when cosmogenic erosion rates are juxtaposed to present day climatic data; they may reflect different climatic conditions (Binnie *et al.* 2008).

highest concentration, which would be the case in an unperturbed profile (Fig. 4.24). Thereby, the average concentration in the mixed layer steadily increases to values that are higher than the mean concentration of a vertically mixed depth–concentration profile (Fig. 4.24). In the steady-state situation, where erosion and vertical perturbation are continuously ongoing, the concentration in the vertical profile approaches the same value that would be expected at the surface of an eroding, unperturbed soil (Eqn. (4.12); Granger and Riebe 2007). Therefore, in basins with steady-state bioturbation and erosion, basin-wide denuation rates obtained with cosmogenic nuclides will reliably reflect the actual rates.

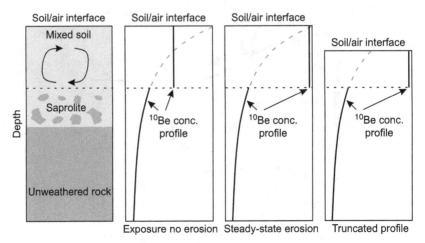

Fig. 4.24. Model predictions for [10]Be depth concentration profiles (black solid lines) through mixed soils. In situations without erosion the mixed section of the profile will reflect the average concentration of the unperturbed profile (stippled grey line). In situations with steady state erosion of the mixed soil layer the mean concentration of the mixed soil layer is equal to the surface concentration of the unperturbed profile (stippled grey line). Truncated soil profiles, for instance truncated by fast erosion triggered by land-use, may still preserve the pre-disturbance erosion rate, if the boundary between mixed layer and unmixed saprolite has not shifted as a consequence of the disturbance. After Granger and Riebe (2007).

Thorough vertical mixing throughout the soil profile increases the method's resilience in situations with recently accelerated modern denudation, e.g. by human activity (Wilkinson and McElroy 2007). As long as the modern erosion has not yet completely removed the mixed layer of a preanthropogenic soil, but just truncated it, the measured cosmogenic erosion rates will reliably reflect natural erosion rates (Granger and Riebe 2007).

The chemical resistance of quartz, as compared to other constituents of soil, can result in its enrichment in soils relative to less-resistant minerals (similar to zircon, see above). In vertically mixed soils, the weathering resistance also increases the average residence time of quartz, as compared to the average soil material. This can lead to the underestimation of steady-state erosion rates by a factor of $1/f_d$ when using the nuclide concentrations in the residual quartz (Small *et al.* 1999, Riebe *et al.* 2001). The *dissolution correction factor* f_d for a soil with thickness h is defined as (Small *et al.* 1999, Riebe *et al.* 2001):

$$f_d = f_e + [1 - f_e]e^{-\rho h/\Lambda} \qquad (4.34)$$

with the enrichment factor f_e given by

$$f_e = \frac{[Zr]_{soil}}{[Zr]_{bedrock}} \qquad (4.35)$$

It is usually assumed that enrichment factors of Zr and quartz are equal (Small *et al.* 1999, Riebe *et al.* 2001, 2004, Granger and Riebe 2007), which is a reasonable assumption in all but the most intensely weathered areas. Naturally, these correction factors are also applicable in situations where local (i.e. not spatially averaged) erosion rates are determined.

Sampling considerations

As with exposure dating, quartz is the main target mineral for studies on local and spatially averaged erosion rates. Its weathering resistance, ubiquitous occurrence and simple target chemistry make it particularly useful (Chapter 2). The majority of studies use [10]Be (cf. von Blanckenburg 2005; Granger and Riebe 2007); however, [21]Ne is utilized for old landscapes (van der Wateren and Dunai 2001, Dunai *et al.* 2005, Kober *et al.* 2007) and for analysing single-grain populations (Codilean *et al.* 2008). [14]C has not yet been used for erosion studies, but would, in principle, lend itself to constrain Holocene erosion rates. Olivine (Gayer *et al.* 2008), and accessory minerals such as zircon, apatite and titanite (Farley *et al.* 2006) can be used to derive erosion rates utilizing cosmogenic [3]He. The choice of target mineral and isotope should be guided by the lithology of the field area, and the temporal resolution required.

Once the target mineral, its grain size and the isotope of choice are known, and the scientific question and sampling locations are defined, the actual sampling is rather straightforward. Avoiding places with evidence or potential for recent disturbance and/or contamination of sediments (human excavations, bank failure, etc.) samples are simply scooped and bagged from the active channel. These samples can be taken from the bottom of the stream bed; alternatively mid-stream sediment bars are useful targets in situations where long-term sediment storage is unlikely (shallow samples of sand-sized sediments deposited below the high-water level). If the above situations are not available/feasible, samples can be taken on the lateral river banks; there particular care has to be taken to avoid situations where old, stored sediments may contaminate the sample. The above strategies can be applied to continuously flowing as well as ephemeral streams and dry washes. If a particular targeted grain size is rare in the sediments, they can be sieved on site to reduce sample weight. In the latter situation the potential grain-size bias on results should be considered (Brown *et al.* 1995; are the rare size fractions representative for the catchment?).

4.3.2 Palaeodenudation rates

Sediments that are buried and subsequently shielded from cosmic rays can be used to derive palaeodenudation rates (van der Wateren and Dunai 2001, Schaller *et al.* 2002, Granger 2006) (see also Section 4.2). This approach requires that the fundamental properties of the palaeo source region can be estimated or reconstructed with reasonable accuracy (e.g. mean elevation), and that postburial production is accounted for (Schaller *et al.* 2002). The more subtle corrections, such as those for dissolution and topographic shielding (see above) usually cannot be applied with confidence. Therefore, palaeodenudation rate determinations will have somewhat higher uncertainties than the corresponding modern steady-state denudation rates. However, due to the uniqueness of information that may be obtained, and the scarcity of information on past process rates, palaeodenudation rate determinations have the potential to fill important gaps in our understanding of the Earth's surficial evolution.

4.4 Uplift rates

The time-integrated cosmic-ray flux and thus the production rates of cosmogenic nuclides are sensitive functions of a site's altitude (Section 1.3). Consequently, the cosmogenic nuclide concentration of a sample inevitably reflects the time-integrated altitude at which it was exposed. In exposure dating (Section 4.1), the altitude of a sample is explicitly assumed to be known and invariant. In some cases, however, uplift (or subsidence) may be significant during or after original exposure. If the exposure history of a sample is suitably constrained, a palaeo-elevation and/or average uplift rate may be derived.

Two scenarios may permit the calculation of a palaeoaltitude and/or uplift rate: (i) exposure for a finite, known exposure period at constant elevation without shielding and erosion, followed by fast/instantaneous burial that occurred at a known time in the past (Blard *et al.* 2005), and (ii) steady-state uplift of a surface with continuous exposure, with no erosion or burial

(Brook *et al.* 1995b, Schäfer *et al.* 1999, Van der Wateren *et al.* 1999, Gosse and Stone 2001, Dunai *et al.* 2005, Riihimaki and Libarkin 2007).

The first scenario may be realized in volcanic deposits, as radiometric ages in a stack of lava flows can constrain the onset and termination of exposure of individual lava flows (Blard *et al.* 2005). The original exposure time of a buried lava flow must be significantly longer than the combined uncertainties of the radiometric age determinations of the two lava flows (i.e. the ages of the previously exposed and of the burying flow) (Blard *et al.* 2005). In practice, this requires preburial exposure durations of at least several tens of thousand years; in some cases up to several hundred thousand years, to obtain a useful resolution of the palaeoaltimetry (Blard *et al.* 2005). This requirement may limit the applicability of this approach. Even in modern field situations with all evidence visible and preserved, it is sometimes a challenge to exclude erosion and/or shielding of lava flows of significant age.

The second scenario, steady-state uplift without erosion or burial, requires long-term stable landforms or extremely fast uplift. At moderately fast uplift rates of $\sim 100\,\mathrm{m\ Ma^{-1}}$, around 1 Ma continuous exposure is required to change the time-integrated production rate at a site by $\sim 5\%$ (relative to the present elevation). This change is similar to the current analytical resolution of the cosmogenic methodology, ignoring the uncertainties in our knowledge of altitude scaling (Chapter 3). Thus, immaculately preserved landforms with ages well in excess of 1 Ma, or young features in extremely fast-uplifting areas ($\sim 10\,\mathrm{km\ Ma^{-1}}$), may potentially be used for this form of palaeoaltimetry. This limits its potential application to ancient landforms in arid and hyper-arid regions as found in Antarctica and the Atacama desert (Schäfer *et al.* 1999, Van der Wateren *et al.* 1999, Dunai *et al.* 2005, Carizzo *et al.* 2008, Evenstar *et al.* 2009) or to active continental margins (Fig. 4.22).

4.5 Soil dynamics

The predictable nature of cosmogenic-nuclide production in the subsurface allows the character and rates of soil processes to be constrained. Any process that affects transport and storage of grains in soils will affect their cosmogenic-nuclide concentration, and the shape of their depth–concentration profiles in the catena. Conceptual models of soil formation and soil dynamics (Minasny *et al.* 2008) should therefore allow qualitative prediction of the spatial distribution (vertically and laterally) of cosmogenic-nuclide concentrations in soils. These predictions can be tested against actually observed patterns, allowing exclusion of alternative models that predict significantly different patterns (Wilkinson and Humphreys 2005).

Due to the multitude and potential complexity of different soil processes, there is no generic approach for using cosmogenic nuclides; each question requires a tailored sampling strategy and analytical approach. The following few examples serve as a proxy for the diverse possibilities *in situ*-produced cosmogenic nuclides provide for soil sciences. However, not only *in situ*-produced cosmogenic nuclides are used in soil sciences, but also meteoric/atmospheric cosmogenic nuclides (Chapter 2) that are incorporated into soils. Where applications utilize meteoric cosmogenic nuclides, this is specifically indicated, and the reader is referred to the corresponding references for the underlying methodology, which is not covered in this book.

Soil production

Systematic variations of cosmogenic-nuclide concentrations in soil-mantled bedrock have successfully been used to verify longstanding but hitherto untested concepts of the soil production function (Gilbert 1877, Heimsath *et al.* 1997, Heimsath *et al.* 1999, 2001, Humphreys and Wilkinson 2007). These studies (Heimsath *et al.* 1997, Heimsath *et al.* 1999, 2001) confirmed the notion that soil production rates in equilibrium landscapes decrease with increasing soil depth. Subsequently, it was demonstrated that the decrease of chemical weathering rates with increasing soil depth is at least partially responsible for that effect, and that bedrock weathering rates are controlled by topography (Burke *et al.* 2007).

Cosmogenic-nuclide depth profiles have been used successfully to investigate the dynamics of laterite (a ferralsol) formation in various tropical environments (Braucher *et al.* 1998a, Braucher *et al.* 1998b, Braucher *et al.* 1998c, Braucher *et al.* 2000).

Material transport

Cosmogenic-nuclide depth profiles in soils from hill slopes indicated that commonly used linear-diffusion equations describing soil transport are only appropriate for slopes with gentle gradient and convex-up shape (Heimsath *et al.* 2005); a depth-dependent transport law was suggested to be more generally applicable instead (Heimsath *et al.* 2005).

Transport by soil creep (Heimsath *et al.* 2002) and along desert piedmonts (Nichols *et al.* 2007) are other examples where cosmogenic nuclides have been used successfully to describe soil transport.

Very short-term soil transport (several days) can be investigated with meteoric [7]Be (Kaste *et al.* 2007).

Soil mixing

Cosmogenic-nuclide depth profiles in soils can be used to detect and quantify the extent of soil mixing from cryoturbation, bioturbation and anthropogenic perturbation (Granger and Riebe 2007, Schaller *et al.* 2009) (Fig. 4.24).

Soil inflation/desert pavements

Stone pavements are a common feature in desert environments and are often interpreted as lag deposits or as the product of continuous soil activity bringing new clasts to the surface (Wells *et al.* 1995). Exposure ages of clasts of desert pavements should therefore be highly variable. This notion was tested by Wells *et al.* (1995) in the Mojave Desert, who found concordant exposure ages of desert-pavement clasts and adjacent uncovered bedrock of the same lithology instead. This finding led to the development of a new model for pavement evolution, which predicts that pavement clasts are continuously exposed at the surface as a consequence of deposition and pedogenic modification of windblown dust (Wells *et al.* 1995).

Authigenic mineral formation

Exposure ages of landforms can place the temporal framework for authigenic mineral formation (Schroeder *et al.* 2001). The formation of pedogenic carbonate in calcretes can be directly dated utilizing ^{36}Cl-exposure dating (Liu *et al.* 1994b). The meteoric component adsorbed onto, or incorporated in, authigenic minerals may be the dominant source of cosmogenic nuclides for some minerals. Studies dating pedogenic carbonate (Liu *et al.* 1994b) or gibbsite growth in soils (Schroeder *et al.* 2001) rely partially on meteoritic cosmogenic nuclides (^{36}Cl and ^{14}C, respectively). It is also possible to use the decay of meteoric ^{10}Be to derive age models for authigenic mineral formation (Barg *et al.* 1997, Lebatard *et al.* 2008).

4.6 Dealing with uncertainty

As with any analytical technique that is applied to scientific questions in nature, the application of cosmogenic nuclides has to cope with various sources of uncertainty. For cosmogenic nuclides, the uncertainty can be split into three sources: (i) analytical/observational, (ii) methodological and (iii) geological/natural. All three sources can harbour systematic and random uncertainties; random uncertainties affect the precision, systematic uncertainties, the accuracy of results (Taylor 1997). To assess the validity and robustness of interpretations that are based on cosmogenic-nuclide measurements, the uncertainties must be characterized. Depending on the application, some interpretive models can be robust, even in the face of large uncertainties. However, other potential applications may be unfeasible at only moderate or small uncertainties. To avoid disappointments, and potentially wasting resources, it is worthwhile to assess the accuracy and precision required to solve a scientific question with the magnitude of anticipated uncertainties, and the efforts that would be required to reduce them. This assessment should ideally occur early in the planning stage of a project. A list of the different sources of error pertinent to cosmogenic studies is given in Table 4.1.

Analytical and observational uncertainty

All observed and estimated quantities describing a sample and the sampling site, relating to corrections for mass shielding (Section 4.1.2) have uncertainties. Examples of observed quantities subject to uncertainties include sample thickness, surface geometry, topographic shielding, elevation, erosion rate and style, and mass shielding by snow, soil, ash etc. For exposure dating, the uncertainty associated with erosion, and its correction, will, in many instances, dominate the overall uncertainty. Corrections for erosion can be large, and uncertainties may be of a magnitude similar to the correction itself. Corrections based on other parameters are usually small. Consequently, uncertainties arising from these corrections are equally small (at most, the same magnitude as the correction itself). Errors of measured sample parameters are mostly random, while estimated parameters are prone to systematic errors (of judgement).

Uncertainties pertaining to basin-averaged erosion rates relate to the violation of conditions 1–6 (Section 4.3.1). Observational uncertainties mainly concern the representativeness of samples (lithology, grain size).

Uncertainties associated with sample preparation and analyses are listed in Table 4.1. Depending on the available equipment, the experience of the operator, and the effort invested (e.g. integration time of AMS) their cumulative uncertainty can be as low as 1–2%; commonly they are somewhat larger, but usually do not exceed 5% for routine samples. The nature of individual preparation and analytical uncertainties are comprehensively discussed by Gosse and Phillips (2001). The overall precision of the preparation and analytical effort is probably most realistically

Table 4.1. *Sources of uncertainties when using cosmogenic nuclides for dating purposes. It is anticipated that the systematic errors on the methodological parameters (*) will be greatly reduced after the research of the international CRONUS consortia is concluded, leaving random error as the dominant source of uncertainties. Methodological uncertainties vary with the nuclide used, and the latitude and altitude of the sampling site. If there is (are) a reliable local calibration site(s) available, methodological uncertainties can be greatly reduced.*

Analytical/observational (2–6%)

Sample parameters (<1–5%)	
Elevation	Random
Surface geometry (self-shielding)	Random
Sample thickness	Random
Thickness and density of overburden (soil, rock)	Random
Preparation (1–3%)	
Cross-contamination	Random/systematic
Sample weighing	Random
Carrier weighing	Random
Dilution of target mineral by non-target minerals	Random/systematic
Stable isotope measurement (e.g. ^{27}Al for ^{26}Al)	Random
Target element chemistry	Random
Major and trace elements (for neutron flux calculations, ^{36}Cl; ^{3}He)	Random
AMS or NG-MS (1–5%)	
Counting statistics (standards and samples)	Random
Machine background correction	Random
Blank correction	Random
Correction for non-cosmogenic components	Random
Characterization of standard material	Random/systematic*

Methodological (5–15%)*

Radionuclide half-life	Random/systematic*
SLHL-production rate	Random/systematic*
Cosmic ray flux attenuation in rocks	Random/systematic*
Scaling factors	Random/systematic*
Angular dependency of cosmic ray flux in the atmosphere	Random/systematic*
Secular variation of atmospheric pressure, magnetic field, solar modulation	Random/systematic*

Geological/natural (0–100%)

Assumed erosion rate and style (for age-correction)	Systematic
Shielding correction (snow, vegetation etc.)	Systematic/random
Pre-exposure	Random
Exhumation	Random
Misjudgment of the geomorphologic context of a surface	Systematic

assessed by the reproducibility of mineral standards (i.e. homogenous material that undergoes the same preparation steps as actual samples, aliquots of which are repeatedly processed alongside samples), or replicate measurements of samples of a particular study (this may be beneficial if samples are analytically challenging, e.g. having near-background concentrations). There are currently no certified mineral standards available, but a homogenous in-house laboratory material would serve the purpose of assessing the precision of routine analyses (Dunai and Stuart 2009; Appendix B). If such material is shared between various preparation laboratories that use different AMS facilities, the accuracy can be gauged as well (Dunai and Stuart 2009). The reproducibility of AMS measurements is already routinely assessed by the measurement of AMS standards. These do not, however, include the uncertainty of the sample preparation and chemical analytical procedures (a fact that is often overlooked in practice). The reproducibility of AMS measurements is the minimum random uncertainty of any sample (internal uncertainty).

Methodological uncertainties

There are four main sources of methodological uncertainties; these are production rates, scaling methods (Chapter 3), half-lives and parameters describing mass attenuation (Section 4.1.2).

It is evident that the uncertainty in production rates of cosmogenic nuclides translates directly into a corresponding uncertainty of the exposure age; ages cannot be known more accurately than the underlying production rates. This external uncertainty is commonly larger (e.g. currently $\sim$10% for ^{10}Be (Balco *et al.* 2008)) than the internal uncertainty of ages (1–5% analytical uncertainty, see above). Approaches for treating the difference between internal and external uncertainties in practice are discussed at the end of this section. The uncertainty of the local production rate that is used for a particular sampling site is intrinsically linked to the uncertainties associated with scaling factors, unless a local calibration site is available (see below).

The uncertainties of scaling methods are, in part, random, i.e. defined by the goodness of fit to proxy data that is used to construct scaling factors (Chapter 3). These can range between 2% at sea level to >10% at high mountain altitudes (>5000 m; e.g. Dunai 2000, 2001); or estimated at $\sim$10% throughout (Lal 1991). Moreover, scaling factors are prone to systematic uncertainties, as they are derived by proxy measurements and use modern data that are extrapolated to the past (Chapter 3). Further, the choice of an inaccurate pressure–altitude relationship can also result in systematically offset production rates and ages (Dunai 2000, Stone 2000, Staiger *et al.* 2007), as can the choice of geomagnetic records underlying the scaling factors (Section 3.3.2). At the time of writing, the testing of currently available scaling factors is still ongoing. An emerging consensus is that the goodness of fit between predicted and measured production rates will be used to assign the uncertainty of scaling factors, encompassing random and systematic uncertainties. This approach is currently limited by the significant scatter in the historic calibration data set available, where, for instance, a single site (Fig. 4.25) dominates calculated global uncertainties (Balco *et al.* 2008); a situation that is being improved by ongoing calibrations of the international CRONUS research consortia. For the time being, it may be claimed that for sampling sites that are at a similar altitude and latitude range to the main body of calibration data (30–45° latitude, >1500 m), scaling uncertainties will be smaller than for samples outside that range (Gosse and Phillips 2001). If there are reliable production-rate calibrations in the geographic vicinity of a field site investigated, uncertainties introduced by scaling factors are negligible (they cancel each other out; Balco *et al.* 2008). In all other situations, uncertainties in scaling factors have a linear effect on calculated ages and rates. For example, a 5% uncertainty in the scaling factor adds 5% to the uncertainty of an age. It is anticipated that ongoing calibration measurements in the framework of CRONUS will reduce the combined external uncertainty of scaling factors and production rates to <5% for commonly used nuclides such as ^{10}Be and ^{26}Al.

The experimentally derived half-lives of cosmogenic radionuclides have random uncertainties in the order of several % (Chapter 2). Commensurate with the timescale of the application, the uncertainty of the half-life will feed through to the final result (decay modifying the nuclide concentrations in buried or exposed samples). In the case of ^{10}Be, alternative propositions for half-lives differ by more than $\pm 2\,\sigma$ (random experimental uncertainty; Section 2.3.1). This indicates that at least one set of values is inaccurate, and would yield systematically wrong results if used in cosmogenic applications. For young samples (exposure time $<<T_{1/2}$) this effect is negligible (<1%). For old samples, (exposure time $\sim T_{1/2}$) though, the error is close to 10% (Nishiizumi *et al.* 2007). At the time of writing, the discussion on the ^{10}Be half-life is nearing a conclusion (Section 2.3.1). Thus, it may be anticipated that in future half-life uncertainties can be considered to be of random nature only.

The exponent of the angular dependency of the cosmic-ray flux through the atmosphere, which is used for topographic shielding

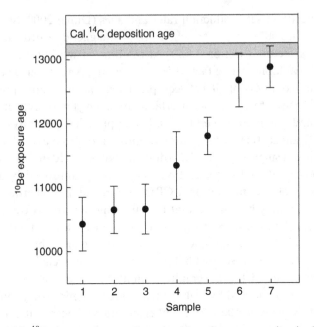

Fig. 4.25. [10]Be exposure ages from boulders from a moraine in Peru (Farber *et al.* 2005). This is a typically sized data set showing significant scatter. Four out of seven data overlap within ±2 σ (±1σ, shown), and could in principle be used to calculate a mean ages (one high age or one low age; with two or three 'outliers' each). However, field-evidence suggests degradation of the moraine by erosion (Farber *et al.* 2005). External age constraints indicate that the oldest samples are nearest to the true age and that younger samples are indeed affected by erosion (Farber *et al.* 2005). The moraine age is probably best represented by the oldest boulders (Putkonen and Swanson 2003); in the original study a mean of the four oldest boulders was advocated (Farber *et al.* 2005). The mean concentration of all boulders is used in the global calibration data set of Balco *et al.* (2008); this single site largely controls the (large) external uncertainty of the global data set (Balco *et al.* 2008). Ongoing measurements within the CRONUS networks should improve this situation in the near future.

corrections (Eqns. (4.14) and (4.15)), has a considerable uncertainty, and the currently most commonly used value of 2.3 may be systematically offset from the actual value for low elevations (Sections 1.3 and 4.1.2). Fortunately, for most sampling situations, topographic shielding corrections are small (<5%), as are the effects of the correction factor uncertainty (e.g. if the correction were 20% uncertain, it would have a 1% effect on a calculated age in a situation with 5% topographic shielding).

In any situation with significant mass shielding (Section 4.1.2), the uncertainty of the mass-attenuation coefficient used (Section 1.4) will contribute to the uncertainty for the mass-shielding correction (Section 4.1.2) However, the uncertainties of the parameters describing the mass shielding, such as thickness, density and duration, will most likely dominate overall uncertainties in this category.

Geological/natural uncertainties

Geological uncertainties arise mostly from our intrinsically limited a priori knowledge of the exposure history of individual samples (pre-exposure, burial) and the uncertainty in the shielding history of a sampling site (snow, soil, ash, vegetation cover).

For instance, in most cases we do not know a priori whether a boulder on a glacial moraine has had significant pre-exposure, whether it has been recently exhumed, or whether it has been eroded (Section 4.1.3). We can tune the sampling strategy to reduce the likelihood and magnitude of effects by exhumation such as by moraine degradation (Section 4.1.3), and in some cases can exclude boulder erosion (Section 4.1.3). However, in most cases, we are unable to categorically exclude pre-exposure. If intermittent snow cover is likely, we would want to correct for it by estimating its average thickness and duration of cover over the entire exposure period (a challenge for climate modellers). Alternatively we might choose large boulders towering over the surrounding landscape features, which are unlikely to have been covered by significant amounts of snow (Section 4.1.3). Even our best attempts at obtaining samples unaffected by pre-exposure, erosion and cover will fail sometimes, and a priori we do not know when we fail (otherwise we would not have sampled the site in the first place). Therefore, data sets such as exposure ages from a single moraine may have significant scatter, and/or significant outliers (see below), despite our best efforts (Figs. 4.25 and 4.26).

In cases where there are suspected outliers, they may be removed using robust statistical criteria. Examples are the commonly used Chauvenet's criterion (Taylor 1997), or the more general, but hitherto less-frequently used, Peirce's criterion (Gould 1855, Ross 2003), which is specifically designed to consider cases with more than one outlier (Taylor 1997, Ross 2003). Alternatively, analogous to procedures used in Ar–Ar and K–Ar dating (Dalrymple and Lanphere 1969, Koppers *et al.* 2000), mean ages can be calculated using exposure ages that fall within $\pm 1.96\sigma$ in the $\pm 1.96\sqrt{\sigma_1^2 + \sigma_2^2}$ confidence envelope of an age population, where σ_1 and σ_2 are the standard deviations (i.e. cumulative random errors) of

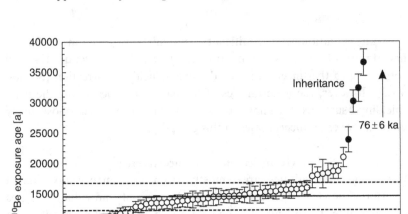

Fig. 4.26. [10]Be exposure ages of the Pommerian moraine, Europe (Rinterknecht *et al.* 2006). This is an exceptionally large data set from a single geomorphological feature, showing nicely the features that can be expected from a normally distributed sample set (white circles) as well as outliers (black circles) due to inheritance or erosion/shielding. The solid line indicates the mean and the stippled lines the 1 σ error envelope of the data. Using any of the methods discussed in the text, the outliers could be identified as such. In the original study several additional samples were excluded using geological considerations. Data from Rinterknecht *et al.* (2006), format of diagram from Granger and Riebe (2007).

the lowest and highest age in the age population, respectively (i.e. in this case the population excluding the suspected outlier(s)). Commonly, the rounded criterion ±2σ is stated for convenience; ±1.96σ denotes the 95% confidence interval (Taylor 1997).

The above weeding methods are applicable to at least three or more observations (ages). There are several other methods to exclude outliers using statistical criteria (ASTM 2008), and it remains the responsibility of the researcher to choose the most suitable method. There is currently no generally accepted procedure within the cosmogenic community to exclude outliers, some practitioners abhor excluding any data. The purist notion against any outlier treatment, however, is somewhat at odds with standard practice in science and engineering (ASTM 2008).

Once outliers, if present, are excluded, and the remaining data overlap within their uncertainties (for isotopic methods customarily generously defined as ±2σ, see above), reliable mean ages can be calculated. These may be the arithmetic mean with a standard deviation of the mean (Taylor 1997) of

$$\sigma_{\bar{x}} = \sigma_x / \sqrt{N} \tag{4.36}$$

with σ_x and N as characteristic uncertainty of individual ages and number of ages, respectively (Taylor 1997). Alternatively, if the individual uncertainties are variable and not similar to the above case, error-weighted means should be calculated instead (Taylor 1997):

$$\text{age}_{\text{mean}} = \frac{\sum 1/\sigma_i^2 \cdot \text{age}_i}{\sum 1/\sigma_i^2} \tag{4.37}$$

The uncertainty σ_{mean} of the error-weighted mean age_{mean} is given by:

$$\sigma_{\text{mean}} = \frac{1}{\sqrt{\sum 1/\sigma_i^2}} \tag{4.38}$$

Sometimes $1/\sigma$ is used as a weighting factor instead of $1/\sigma^2$; the former gives more weight to samples with large errors relative to the latter.

If individual exposure ages of a group of ages from a landscape feature do not overlap within ±2σ, this can either be taken as indication that the random errors are seriously underestimated (Taylor 1997), or that some/all calculated ages are affected by pre-exposure and/or erosion/shielding (non-random), and thus do not describe a unique age population. Sometimes a mean age is nevertheless calculated from such a scattered population. It would be wrong to calculate the standard deviation of the mean (Eqn. (4.36)) or of an error-weighted mean (Eqns. (4.37) and (4.38)) for such a population (Taylor 1997), as they would seriously underestimate the uncertainty (because for such populations, the individual uncertainties are evidently unrelated to the population's variance). The variance of the mean age should be used instead.

If there are less than three overlapping ages (within ±2σ), and more outliers than 'good' ages, it is not very likely that the mean age of a scattered population accurately describes the actual age of the feature (Fig. 4.25). Deviations of this rule of thumb are possible, and should be revealed/confirmed by external evidence (e.g. field evidence, independent age constraints). For instance, in cases where erosion and/or mass shielding are the most likely/only culprits for scatter (e.g. lava flows), the oldest ages can be safely interpreted as minimum ages, and if several ages define a sharp upper limit, they are likely to represent the exposure age of such a surface.

The scatter of any exposure-age population itself can be used to assess the magnitude of the combined effects of pre-exposure and erosion/shielding, assuming they affect samples randomly. Their combined effects should not be larger than the observed scatter, unless erosion obliterated the inherited memory of the original surface. The scatter also includes random analytical uncertainties. Therefore, the variance of a normally distributed age population can be used to constrain the maximum combined analytical uncertainty (Gosse and Phillips 2001).

Error propagation

The above discussion often remains relatively vague concerning the size of individual uncertainties, not least because most are estimates. Some quantities, such as the cumulative analytical uncertainty, can be directly assessed, e.g. by replicate measurements. Methodological and geological uncertainties, on the other hand, are probably significant, but are loosely constrained, the former because of yet to be concluded renovation of the technique (CRONUS), the latter due to intrinsic limitations. If conservative error estimates are fully propagated, the resulting cumulative errors would probably indicate rather unreliable results, which are, however, not common experience when exposure ages are compared to independent age constraints (Gosse and Phillips 2001). Thus, assigning maximum uncertainties to all parameters is pointless, as it will underestimate the accuracy of the technique (Gosse and Phillips 2001). Further, not all parameters and their uncertainties are independent from each other. Thus, errors are not always additive (Gosse and Phillips 2001). The fact that probably the largest uncertainties are systematic and arise from methodology (production rates, half-life, scaling, pressure–altitude relationship) permits a differential treatment of data.

The concentrations of a given cosmogenic nuclide in samples from one area will be affected to the same extent by methodological uncertainties. Thus, when comparing results within one area, e.g. to establish a relative chronology of events, the methodological uncertainties (external uncertainties, Balco *et al.* 2008) do not need to be considered. This essentially reduces the uncertainty to the analytical and geological uncertainties, i.e. the internal uncertainty (scatter) of the data set.

However, if results of different nuclides are compared, or if the cosmogenic chronology is compared to results from different regions or to independent age records, the relevant methodological uncertainties need to be considered (Gosse and Phillips 2001, Balco *et al.* 2008). In these situations, comparison against local independent age constraints and/or using multiple nuclides can help to assess the magnitude and presence of any systematic offset (Gosse and Phillips 2001).

Appendix A

Sampling checklist

The sampling checklist and field-note template are adapted/amended from the protocol developed at the CRONUS-Earth Workshop on Sampling Protocols and Calibration Sites, Coeur d'Alene, Idaho, 26–28 May 2005 (Nat Lifton, personal communication).

Each sample site should be carefully described for potential problems that may affect the results. A list of the potential problems should be compiled for each site and critically discussed, with an assessment of the risk of these issues having influenced the exposure history. Checklists of potential site issues that should be discussed if applicable might include:

1. Shielding (past and present). Include evidence of, and estimated potential effects of each:
   - Ash cover
   - Soil cover
   - Snow cover
   - Ice/glacier cover
   - Water/lake cover
   - Vegetation
   - Topographic shielding
2. Stability of surfaces to be sampled:
   - Have boulders or surfaces rolled or shifted since deposition? Justify this.
   - Is the surface original? Justify this.
3. Integrity of the surface to be sampled:
   - Evidence of exhumation (e.g. boulder exhumed from a moraine)?
   - Evidence of erosion (e.g. boulder erosion)?
4. Risk assessment of inheritance:
   - Could the surface have a complex exposure history?
   - If possible this should be quantitatively assessed by sampling both shielded and unshielded samples at the same site.

The following data should be recorded in the field:

1. Date and time
2. GPS latitude and longitude, elevation

144

3. Logistics details: travel time, distances to road, etc.
4. Sketch of the location
5. Photographs both near- and far-field from different angles, with photograph azimuth and sun angle noted. Photographs should include an easily visible metric scale. Sample locations should be photographed both before and after sample collection
6. Measurements of the site geometry, including:
   - Description of the location sampled
   - Shape and dimensions of outcrop, boulder or lava flow features. All principle dimensions should be measured and recorded
   - Is the feature typical/unique at this site?
   - Measurements of sampling location on the outcrop or boulder
   - Sampled rock thickness
   - Topographic shielding
   - Surface dip and dip direction
   - Sample lithology: rock type, estimated mineral abundances and grain-sizes, visible cracks and voids, veins and other heterogeneities
7. Visual description of appearance of the sample:
   - Colour
   - Lichens
   - Weathering characteristics: mineral discoloration, staining and varnish formation, differential weathering (e.g. emergent veins, cavernous weathering), ventifacts, top–bottom contrast on clasts, etc.

Further sampling considerations

When sampling boulders or bedrock surfaces, ideal sampling locations are the centres of level surfaces ($<30°$ dip) that are free of cracks or other features that might indicate preferential erosion in the past. In practice, surfaces that are ideal for exposure dating, due to their long-term resistance to weathering and erosion, are also rather resistant to our sampling efforts. Rock chisels with tungsten-carbide tips, core drills and rock saws often provide good sampling results, even on unweathered, smooth surfaces. Sampling with rock saws and core drills has the added advantage that it is relatively easy to obtain samples of an approximately constant thickness. The disadvantage of using these tools is that they leave highly visible and permanent scars on surfaces. In regions that are visited by the public and/or are of scenic beauty, it may be sensible to avoid rock-saw or drill sampling. The scars left by chip sampling as obtained with chisels, have a more natural appearance, and will, in many sampling situations, blend in with the unscarred surface on the timescale of several years/decades. The inevitably lower accuracy of the sample thickness obtained by chip-sampling will have a 'price-tag' of less than $\pm 1\%$ on the final result in most cases (assuming an average sample thickness of $\sim 4\,\mathrm{cm}$, and an uncertainty of $\pm 1\,\mathrm{cm}$); this may well be a 'price' worth paying to maintain the acceptance of the public (or park rangers).

Template for sampling notes

Sample number: Site:
Collector name: Date:

Before Collection:

Describe the condition of the surface (erosion, striations, etc.). Include comments
on possible disturbances, biological condition near site, and field observations on
inheritance or other important factors (see checklist above):

Surface dip/dip direction _____
Lithology _____
Latitude _____ Longitude _____
Datum _____
Elevation_____ Map used_____
General boulder shape (include in sketch):_____
Boulder/rock measurements: x-axis _____ y-axis _____
z-axis _____
Other pertinent length (specify) _____
Sketch rock (if desired):

Photographs of whole rock before sampling:

Looking towards:_____ Camera Initials_____ Photo Number _____
Looking towards:_____ Camera Initials_____ Photo Number _____
Looking towards:_____ Camera Initials_____ Photo Number _____
Looking towards:_____ Camera Initials_____ Photo Number _____

Distance

Looking towards:_____ Camera Initials_____ Photo Number _____
Looking towards:_____ Camera Initials_____ Photo Number _____
Looking towards:_____ Camera Initials_____ Photo Number _____
Looking towards:_____ Camera Initials_____ Photo Number _____

While sampling

Looking towards:_____ Camera Initials_____ Photo Number _____

Other photographs:

Looking towards:_____ Camera Initials_____ Photo Number _____
Looking towards:_____ Camera Initials_____ Photo Number _____

After Collection:

Sample Thickness_____   Estimated Mass_____
Sample Length _____   Sample Width _____

Azimuth   Inclination above horizon

——————— ———————   ——————— ———————
——————— ———————   ——————— ———————
——————— ———————   ——————— ———————
——————— ———————   ——————— ———————
——————— ———————   ——————— ———————
——————— ———————   ——————— ———————
——————— ———————   ——————— ———————
——————— ———————   ——————— ———————
——————— ———————   ——————— ———————
——————— ———————   ——————— ———————
——————— ———————   ——————— ———————

If horizon picture taken, record information here:

Camera Initials_____ Photo Number_____

Photographs of remaining rock after sampling:

Looking towards:_____ Camera Initials_____ Photo Number _____
Looking towards:_____ Camera Initials_____ Photo Number _____

Top of sample (show orientation and scale in picture)

Camera Initials_____ Photo Number_____

Profile view of sample

Camera Initials_____ Photo Number_____

Comments for sample preparation or other comments on the sample:

Appendix B

Reporting of cosmogenic-nuclide data for exposure age and erosion rate determinations

Adapted from: Dunai and Stuart (2009) Quaternary Geochronology, doi:10.1016/ j.quageo.2009.1004.1003

B.1 Introduction

The use of *in situ*-produced cosmogenic nuclides has developed into a versatile quantitative tool for studying the Earth's surface. Since 2004, two major international initiatives (CRONUS-EU in Europe and CRONUS-Earth in North America) have worked on refining the methodology, and on achieving a consensus regarding scaling of the cosmic-ray flux. This work is likely to affect the way exposure ages or erosion rates are calculated in the future. Currently, exposure ages and erosion rates are calculated using a variety of protocols (Lal 1991, Dunai 2000, Stone 2000, Dunai 2001a, Desilets and Zreda 2003, Lifton *et al.* 2008), for discussion see Chapter 2. In order to allow the comparison of age and/or rate information from different studies, as well as safeguarding the future value of published cosmogenic data, it is necessary that all the information required to calculate exposure ages and erosion rates is reported. Guidelines are proposed here for publishing *in situ* cosmogenic-nuclide data for exposure-age and erosion-rate determinations. Similar reporting guidelines are used for other dating techniques, such as radiocarbon dating (Stuiver and Polach 1977, Mook and van der Plicht 1999, van der Plicht and Hogg 2006), and more generally for isotope geochemistry (Goldstein *et al.* 2003). These guidelines significantly expand on earlier suggestions on publication of cosmogenic-nuclide data (Gosse *et al.* 1996), reflecting the ongoing refinement of cosmogenic-nuclide methodology.

In the following, where it is stated that certain information 'must' be provided, it is felt that the omission of that information would seriously affect the future usefulness of the data; either by introducing an unresolved ambiguity of 5% or more and/or by reducing the overall credibility of the study. Information that 'should' be provided is necessary to make comparisons at the 5% level, and/or to maintain good scientific practice. Note that the ambiguity introduced by several omissions in reporting is additive, thus best avoided. For the physical background of the necessity to report the relevant site-specific information see Chapters 1–4.

Table B.1. *Site-specific information as is usually derived from field observations/ measurements*

Site data	Comments
Latitude and longitude	With reference datum (e.g. WGS 84)
Elevation	Reported relative to sea level; with uncertainty
Thickness/depth of sample	Measured value
Density of sample/ overburden	Measured or estimated value
Topographic/geometric shielding	State whether applied or not; if yes, specify the value used and the data/parameters used to calculate it
Cover correction	State whether applied or not; if yes, specify the value used and the data/parameters/ assumptions used to calculate it
Erosion correction	State whether applied or not; if yes, specify the value used and evidence/assumptions used to derive it

The required information, which is detailed in the following sections, is summarized in two tables (Tables B.1 and B.2). In many cases the required information will exceed the space and format of the journal the data is published in; in these instances the information should be provided in the supplementary online data repository of the corresponding journal.

B.2 Site-specific information

Latitude, longitude and *elevation* of sampling sites must be provided as essential parameters to calculate the cosmic-ray flux. Altitude should be reported relative to sea level, providing estimated or determined uncertainties. Latitude and longitude information should be accompanied with the reference datum used (preferably WGS-84 when using a GPS; or the map datum when using a topographic map to locate the site). The accuracy of the information provided should allow relocation of the sampling site with confidence (Table B.1).

The *sample thickness* and *density* must be reported (Table B.1), even in cases where no *self-shielding correction* is applied in the original study. If a density determination is impractical or deemed unnecessary, the assumed density of the rock should be provided. If a self-shielding correction is applied, the *effective attenuation length* used must be provided (Table B.2).

Topographic and geometric shielding records the obstruction of the cosmic-ray flux by the surrounding topography and other nearby obstructions, including the self-shielding on inclined surfaces. All studies should provide the numerical value used to correct for topographic shielding of the sampling site or provide a statement that no shielding correction has been applied. In cases where the effect

Table B.2. *Essential data and information required to evaluate exposure ages and erosion rates calculated from cosmogenic-nuclide data*

Data/Information	Comments
Generic information:	
Concentration	Report concentrations; provide information on reference standard material and half-lives used. Identify/characterize any mineral standards used (if applicable).
Blank correction	Report blank value(s); state whether a correction is applied.
Interference correction	Any corrections for contributions from non-cosmogenic sources of the nuclide analysed need to be described; uncorrected values should be reported alongside any corrected values.
Analytical error	State which uncertainties are included in calculation and whether 1σ or 2σ errors are reported
Nuclide-specific sample data:	
^{10}Be, ^{26}Al, ^{21}Ne	Specify sample material used; if it is not quartz, report measured major-element composition.
^{36}Cl	Report measured major-element composition and concentrations of neutron-flux-relevant trace elements (Li, B, Cl, Cr, Sm, Gd, U and Th) of whole rock. Report the measured or estimated water content of the whole rock. If mineral separates or leached whole rock samples are used, provide, *additionally*, the major-element composition of these materials or, as a bare minimum, provide the target-element concentrations (K, Ca, Fe, Ti and Cl).
^{3}He	Report major-element composition and Li, U, and Th concentrations of the target mineral along with information on the shape and size the target mineral had in the host rock. State whether the outermost 20–30 microns have been removed in the course of sample preparation; if they have not been removed, report Li, U and Th concentrations of the host rock. Provide information on the geological age of the rock.
Information on exposure age and production-rate calculations:	
Cosmic-ray scaling and attenuation	Report scaling scheme and effective subsurface attenuation used.
Production rates	Report the calibration data set(s), and/or target (element)-specific production rates used.
Published online/offline calculators	If a published calculator is used, report its version. State whether default or user-defined parameters were applied; all user-defined parameters must be reported.

Table B.2. (*cont.*)

Data/Information	Comments
Geomagnetic records	State whether geomagnetic corrections have been made; if yes, provide the sources of the geomagnetic records and describe the procedure used *or* report the calculator version, if the geomagnetic correction is intrinsic to the calculator.
Atmospheric pressure	State the pressure–altitude relationship used, *and/or* report the calculator version if this relationship is intrinsic to the calculator used.
Muons	Provide the muogenic contribution to the nuclide production assumed.
Low-energy neutrons	If applicable, describe how the contribution from thermal and epithermal neutron reactions to the cosmogenic inventory was calculated *or* report the calculator version if this calculation is intrinsic to the calculator used.

of shielding on production rate is large (>5%), the measured horizon geometry, i.e. azimuth and zenith angles of the horizon, and/or the slope and strike of the surface sampled should be provided. In situations where shielding is highly significant (e.g. beneath cliffs or near-vertical fault scarps), sufficient detail must be provided to allow the calculation of shielding factors. In all cases where a shielding correction has been applied, it should be stated which angular dependency of cosmic-ray flux was assumed (i.e. which exponent (x) in the relationship: *cosmic-ray flux* $\sim sin^x(\theta)$ was used, with θ being the zenith angle). Commonly $x = 2.3$ is used, but not universally (Chapters 1 and 4). In situations with heavy shielding this information is essential.

Any *field observations* of locally relevant factors that might affect the exposure of the sampling site, such as the likelihood of intermittent cover by snow, soil or volcanic ash, should be reported. If a *cover correction* is performed, the correction factor must be reported. The assumed thickness, density and duration of cover and relevant evidence should be given.

Any relevant information pertaining to the time-integrated erosion affecting a site used for exposure dating should be provided. If an *erosion-rate correction* is applied to the calculation of an exposure age, the actual value of the erosion rate used should be clearly stated. A summary of site-specific information required is provided in Table B.1.

B.3 Cosmogenic-nuclide data

The concentration of the cosmogenic nuclides analysed, including analytical uncertainties, must be provided. It should be clearly stated whether uncertainties reported are at the 1σ or 2σ level (Table B.2). Any corrections for procedural blanks should be described, and relevant blank values provided. In the case of noble-gas isotopes, the measurement of the cosmogenic component often involves correction for atmospheric, magmatic, radiogenic or nucleogenic components (e.g. Kurz 1986a, Niedermann *et al.* 1994, Dunai and Wijbrans 2000, Hetzel *et al.* 2002a, Farley *et al.* 2006, Blard *et al.* 2008b). The correction procedure must be described in the necessary detail, and uncorrected and corrected values should be provided using established procedures and units (Goldstein *et al.* 2003).

The determination of isotopic ratios by mass spectrometry relies upon relating the isotopic ratio of a reference standard material to that of the unknown. Standard materials are rarely universal, and values may need to be recalculated to a common standard in order to allow global comparison. Therefore the standard material used, and the actual or nominal ratios, must be reported (Goldstein *et al.* 2003, Nishiizumi *et al.* 2007). In this context it is also important to state which half-life is assumed for radionuclides. This is of particular importance for ^{10}Be (Nishiizumi *et al.* 2007).

Unfortunately there is still a lack of widely available mineral standards for cosmogenic nuclides that allow the accuracy of the entire process from sample preparation to mass-spectrometric measurement to be assessed. The reference standards mentioned above assess only the performance of the accelerator and the noble-gas mass spectrometers. Several mineral standards are currently in development. Measured ratios of these materials, analysed as unknowns, should be routinely reported as soon as they become generally available. In the meantime, reported values of replicate analysis of an identified in-house standard material, or an identified material that is shared with a group of other laboratories, can serve to assess analytical precision. Further, they serve to peg analytical results to a traceable material that should be linked to international standards when the latter become available.

B.4 Other analytical and observational data

In order to calculate the production rate of the commonly used terrestrial cosmogenic nuclides, and their non-cosmogenic production pathways, it is often necessary to provide supplementary chemical data and/or petrographic information. The provision of chemical data should follow standard guidelines (e.g. Goldstein *et al.* 2003). Most cosmogenic nuclides have specific requirements. A summary of the required nuclide-specific analytical and observational data, which is detailed in the following, is given in Table B.2.

^{10}Be, ^{26}Al and ^{21}Ne are most commonly analysed in quartz. For the purposes of determining the production rates of these nuclides, the chemical composition of quartz is invariant and well defined by its nominal composition. However, if minerals other than quartz are used, the major element composition must be reported. Due to the nature of the production mechanisms for ^{10}Be, ^{26}Al and ^{21}Ne (Gosse and Phillips 2001, Niedermann 2002), the chemical composition of the host rock has only marginal influence on production rates, and therefore does not need to be reported.

^{36}Cl is analysed in minerals and whole rocks with a wide range of compositions. The major element composition of samples must be reported. ^{36}Cl has an important reaction pathway involving thermal-neutron capture by ^{35}Cl. Consequently the concentration of all elements that affect the thermal-neutron flux in a rock need to be reported (i.e. Li, B, Cl, Cr, Sm, Gd, U and Th) (Phillips *et al.* 2001, Schimmelpfennig *et al.* 2009). Further, the water content of the rock must be determined or estimated, and the corresponding values provided. If ^{36}Cl is analysed in mineral separates, then at least the concentrations of target elements (K, Ca, Fe, Ti and Cl) in these minerals must be provided, in addition to the whole-rock composition (Schimmelpfennig *et al.* 2009).

Cosmogenic ^{3}He is increasingly being analysed in a range of He-retentive minerals (Niedermann 2002, Farley *et al.* 2006). The major element composition of these minerals must be reported. There are several different element-specific ^{3}He production-rate factors (Masarik and Reedy 1996, Schäfer *et al.* 1999, Kober *et al.* 2005). Consequently it is necessary to report the scheme used. Further, due to the significant ejection distances of cosmogenic ^{3}He, nucleogenic ^{3}He and radiogenic ^{4}He, the grain size of the minerals analysed, as they were in the rock prior to crushing, should be reported. In cases where the density of the minerals analysed is significantly different from the host rock (e.g. zircon, iron oxides), information on grain dimensions becomes essential (Farley *et al.* 2006, Dunai *et al.* 2007).

The capture of thermalized cosmogenic neutrons by ^{6}Li is an important non-spallogenic source of ^{3}He at the Earth's surface (Dunai *et al.* 2007). Helium-retentive minerals in mafic and ultramafic rocks commonly have low lithium concentrations and therefore low nucleogenic ^{3}He-concentrations. However, He-retentive minerals in evolved rocks commonly have Li-concentrations >5 ppm. In these lithologies, Li-concentration of the surrounding host rock and/or host minerals must be reported, due to the significant ejection distance of tritium, which is a precursor of ^{3}He and is produced in a neutron-capture reaction with ^{6}Li. In cases where large (>600 micron diameter) mineral grains are analysed, or where the outer $\geq$30 micron layer has been removed, the implanted nucleogenic tritium is unimportant, and the Li-concentration of direct neighbours is no longer required (Dunai *et al.* 2007). In all cases where the Li concentration of the mineral analysed is expected to be >5 ppm, the concentration of all elements that affect the thermal-neutron flux in the host rock should be reported (see ^{36}Cl).

In order to correct for implanted and *in situ*-produced radiogenic ^{4}He, which may interfere with the correction for magmatic ^{3}He; the U and Th concentrations of the rock and phenocrysts should be reported (Blard and Farley 2008). Further, information on the geological age of the rock should be provided if it is likely to be older than the exposure age (Blard and Farley, 2008). If the outermost 20 μm of the phenocrysts have been removed, the U and Th concentrations of the host rock are no longer relevant (Blard and Farley 2008).

B.5 Calculation parameters

Besides the analytical and observational data outlined above, all information required for reproducing the calculations of exposure ages and/or erosion rates, as performed by the authors, must be reported. The minimum information that needs to be provided is: (i) scaling factors, (ii) geomagnetic records (if time-dependent scaling is used), (iii) the assumed muogenic contribution, (iv) the altitude–pressure relationship (e.g. standard atmosphere, actual or modelled site pressure) and (v) calibration data sets used. Further, (vi), for nuclides with low-energy neutron reaction pathways, the relevant calculations need to be described/ referenced. If a published calculator is used for calculations (e.g. Phillips and Plummer 1996, Vermeesch 2007, Balco *et al.* 2008) this should be stated, together with the version of the calculator used. If parameters such as geomagnetic records are intrinsic to a calculator (i.e. cannot be changed by the user) reporting the version of the calculator is sufficient. Full information on any additional parameters (i.e. non-default parameters in the case of calculators) used in calcula-tions (e.g. assumed uplift or erosion rates; calibration data sets used) must be provided.

B.6 Concluding note

The guidelines provided above are appropriate for the vast majority of current cosmogenic-nuclide applications to Earth surface sciences. Some specialist appli-cations may need additional information in order to draw the relevant conclusions from the data presented; whenever this is the case, the authors should provide this information. The nature of yet other applications may have as a consequence that authors simply cannot supply all the information outlined above. In this case, a careful argument on the robustness of the data and conclusions, in the light of the 'missing' information, should be presented.

References

Ackert, R. P., Singer, B. S., Guillou, H., Kaplan, M. R. and Kurz, M. D., 2003. Long-term cosmogenic He-3 production rates from Ar-40/Ar-39 and K-Ar dated Patagonian lava flows at 47 degrees S, *Earth Planet. Sci. Lett.* **210** 119–136.

Agnew, H. M., Bright, W. C. and Froman, D., 1947. Distribution of neutrons in the atmosphere, *Phys. Rev.* **72** 203.

Albrecht, A., Herzog, G., Klein, J., Dezfouly-Arjomandy, B. and Goff, F., 1993. Quaternary erosion and cosmic ray exposure history drived from ^{10}Be and ^{26}Al produced *in situ* – An example from Pajarito Plateau, Valles Caldera region, *Geology* **21** 551–554.

Albrecht, A., Schnabel, C., Vogt, S., Xue, S., Herzog, G. F., Begemann, F., Weber, H. W., Middleton, R., Fink, D. and Klein, J., 2000. Light noble gases and cosmogenic radionuclides in Estherville, Budulan, and other mesosiderites: Implications for exposure histories and production rates, *Meteoritics Planet. Sci.* **35** 975–986.

Alvarez-Marron, J., Hetzel, R., Niedermann, S., Menendez, R. and Marquinez, J., 2008. Origin, structure and exposure history of a wave-cut platform more than 1 Ma in age at the coast of northern Spain: A multiple cosmogenic nuclide approach, *Geomorphology* **93** 316–334.

Anderson, R. K., Miller, G. H., Briner, J. P., Lifton, N. A. and DeVogel, S. B., 2008. A millennial perspective on Arctic warming from C-14 in quartz and plants emerging from beneath ice caps, *Geophys. Res. Lett.* **35** L01502.

Anderson, R. S., Repka, J. L. and Dick, G. S., 1996. Explicit treatment of inheritance in dating depositional surfaces using *in situ* ^{10}Be and ^{26}Al, *Geology* **24** 47–51.

Andrews, J. N., 1985. The isotopic composition of radiogenic helium and its use to study groundwater movements in confined aquifers, *Chem. Geol.* **49** 339–351.

Andrews, J. N., Davis, S. N., Fabryka-Martin, J., Fontes, J.-C., Lehmann, B. E., Loosli, H. H., Michelot, J.-L., Moser, H., Smith, B. and Wolf, M., 1989. The in situ production of radioisotopes in rock matrices with particular reference to the Stripa granite, *Geochim. Cosmochim. Acta* **53** 1803–1815.

Andrews, J. N. and Kay, R. L. F., 1982. Natural production of tritium in permeable rocks, *Nature* **298** 361–363.

155

ASTM, 2008. *Standard Practice for Dealing with Outlying Observations*, West Conshohocken: ASTM International, 18 pp.

Audi, G., Bersillon, O., Blachot, J. and Wapstra, A. H., 2003. The NUBASE evaluation of nuclear and decay properties, *Nucl. Phys. A* **729** 3–128.

Badenhoop, J. K. and Weinhold, F., 1997. Natural steric analysis: *Ab initio* van der Waals radii of atoms and ions, *J. Chem. Phys.* **107** 5422–5432.

Baglin, C. M., 2008. Nuclear Data Sheets for A = 81, *Nucl. Data Sheets* **109** 2257–2437.

Bähr, R., Lippolt, H. J. and Wernicke, R. S., 1994. Temperature-induced ^{4}He degassing of specularite and botryoidal hematite: A ^{4}He retentivity study, *J. Geophys. Res.* **99** 17695–17707.

Balco, G., Briner, J., Finkel, R., Rayburn, J. A., Ridge, J. C. and Schaefer, J. M., 2009. Regional beryllium-10 production rate calibration for late-glacial northeastern North America, *Quat. Geochronol.* **4** 93–107.

Balco, G. and Rovey, C. W., 2008. An isochron method for cosmogenic nuclide dating of buried soils and sediments, *Am. J. Sci.* **308** 1083–1114.

Balco, G., Rovey, C. W. and Stone, J. O., 2005. The first glacial maximum in North America, *Science* **307** 222.

Balco, G. and Schaefer, J. M., 2006. Cosmogenic-nuclide and varve chronologies for the deglaciation of southern New England, *Quat. Geochronol.* **1** 15–28.

Balco, G. and Shuster, D. L., 2009. Production rate of cosmogenic ^{21}Ne in quartz estimated from ^{10}Be, ^{26}Al, and ^{21}Ne concentrations in slowly eroding Antarctic bedrock surfaces, *Earth Planet. Sci. Lett.* **281** 48–58.

Balco, G., Stone, J. O., Lifton, N. A. and Dunai, T. J., 2008. A complete and easily accessible means of calculating surface exposure ages or erosion rates from ^{10}Be and ^{26}Al measurements, *Quat. Geochronol.* **3** 174–195.

Ballentine, C. J. and Burnard, P. G., 2002. Production, release and transport of noble gases in the continental crust. In: D. Porcelli, C. J. Ballentine and R. Wieler, (Eds), *Noble Gases in Geochemistry and Cosmochemistry*, Reviews in Mineralogy and Geochemistry 47, Washington: The Mineralogical Society of America, pp. 481–538.

Barford, N. C. and Davis, G., 1952. The angular distribution and attenuation of the star-producing component of cosmic rays, *Proc. Royal Soc. London A* **214** 225–237.

Barg, E., Lal, D., Pavich, M. J., Caffee, M. and Southon, J. R., 1997. Beryllium geochemistry in soils: evaluation of Be-10/Be-9 ratios in authigenic minerals as a basis for age models, *Chem. Geol.* **140** 237–258.

Barker, D. L., Jull, A. J. T. and Donahue, D. J., 1985. Excess ^{14}C abundances in uranium ores – possible evidence for emission from uranium-series isotopes, *Geophys. Res. Lett.* **12** 737–740.

Benedetti, L., Finkel, R., King, G., Papanastassiou, D., Ryerson, F., Flerit, F., Farber, D. and Stavrakakis, G., 2003. Motion on the Kaparelli fault (Greece) prior to the 1981 earthquake sequence determined from ^{36}Cl cosmogenic dating, *Terra Nova* **15** 118–124.

Benedetti, L., Finkel, R., Papanastassiou, D., King, G., Armijo, R., Ryerson, F. J., Farber, D. and Flerit, F., 2002. Postglacial slip history of the Sparta fault (Greece) determined by ^{36}Cl cosmogenic dating: evidence for non-periodic earthquakes, *Geophys. Res. Lett.* **29** 8701–8704.

Bernatorwicz, T., Brannon, J., Cowsik, R., Hohenberg, C. and Podosek, F. A., 1993. Precise determination of relative and absolute β-decay rates of ^{128}Te and ^{130}Te, *Phys. Rev. C* **47** 806–825.

Bhattacharyya, A. and Mitra, B., 1997. Changes in cosmic ray cut-off rigidities due to secular variations of the geomagnetic field, *Ann. Geophys.* **15** 734–739.

Bierman, P. and Steig, E. J., 1996. Estimating rates of denudation using cosmogenic isotope abundances in sediment, *Earth Surf. Landforms* **21** 125–139.

Bierman, P. R. and Caffee, M., 2001. Slow rates of rock surface erosion and sediment production across the Namib Desert and escarpment, southern Africa, *Am. J. Sci.* **301** 326–358.

Bierman, P. R., Gillespie, A. R., Caffee, M. W. and Elmore, D., 1995. Estimating erosion rates and exposure ages with ^{36}Cl produced by neutron activation, *Geochim. Cosmochim. Acta* **59** 3779–3798.

Binnie, S. A., Phillips, W. M., Summerfield, M. A. and Fifield, L. K., 2006. Sediment mixing rapidly and basin-wide cosmogenic nuclide analysis in eroding mountainous environments, *Quat. Geochronol.* **1** 4–14.

Binnie, S. A., Phillips, W. M., Summerfield, M. A. and Fifield, L. K., 2007. Tectonic uplift, threshold hillslopes, and denudation rates in a developing mountain range, *Geology* **35** 743–746.

Binnie, S. A., Phillips, W. M., Summerfield, M. A., Fifield, L. K. and Spotila, J. A., 2008. Patterns of denudation through time in the San Bernardino Mountains, California: Implications for early-stage orogenesis, *Earth Planet. Sci. Lett.* **278** 62–72.

Blard, P. H., Bourlès, D., Pik, R. and Lave, J., 2008a. In situ cosmogenic Be-10 in olivines and pyroxenes, *Quat. Geochronol.* **3** 196–205.

Blard, P. H. and Farley, K. A., 2008. The influence of radiogenic ^{4}He on cosmogenic ^{3}He determinations in volcanic olivine and pyroxene, *Earth Planet. Sci. Lett.* **276** 20–29.

Blard, P. H., Lave, J., Pik, R., Quindelleur, X., Bourles, D. L. and Kieffer, G., 2005. Fossil cosmogenic ^{3}He record from K–Ar dated basaltic flows of Mount Etna volcano (Sicily, 38°N): Evaluation of a new paleoaltimeter, *Earth Planet. Sci. Lett.* **236** 613–631.

Blard, P. H., Pik, R., Lavé, J., Bourlès, D., Bunard, P. G., Yokochi, R., Marty, B. and Trusdell, F., 2006. Cosmogenic ^{3}He production rates revisited from evidences of grain size dependent release of matrix sited helium, *Earth Planet. Sci. Lett.* **247** 222–234.

Blard, P. H., Puchol, N. and Farley, K. A., 2008b. Constraints on the loss of matrix-sited helium during vacuum crushing of mafic phenocrysts, *Geochim. Cosmochim. Acta* **72** 3788–3803.

Blau, M. and Wambacher, H., 1937. Disintegration processes by cosmic-rays with simultaneous emission of several heavy particles, *Nature* **140** 585.

Boaretto, E., Berkovits, D., Hass, M., Hui, S., Kaufman, A., Paul, M. and Weiner, S., 2000. Dating of prehistoric caves sediments and flints using ^{10}Be and ^{26}Al in quartz from Tabun Cave (Israel), *Nucl. Instr. Meth. Phys. Res. B* **172** 767–771.

Boezio, M., Carlson, P., Francke, T., Weber, N., Suffert, M., Hof, M., Menn, W., Simon, M., Stephens, S. A., Bellotti, R., Cafagna, F., Circella, M., De Marzo, C., Finetti, N., Papini, P., Piccardi, S., Spillantini, P., Ricci, M., Casolino, M.,

De Pascala, M. P., Morselli, A., Picozza, P., Sparvoli, R., Barbiellini, G., Schiavon, P., Vacchi, A., Zampa, N., Grimani, C., Mitchel, J. W., Ormes, J. F., Streitmatter, R. E., Bravar, U., Golden, R. L. and Stochaj, S. J., 2000. Measurement of the flux of atmospheric muons with the CAPRICE94 apparatus, *Phys. Rev. D* **63** 032007.

Bramblett, R. L., Ewing, R. I. and Bonner, T. W., 1960. A new type of neutron spectrometer, *Nucl. Instr. Meth. Phys. Res.* **9** 1–12.

Braucher, R., Benedetti, L., Bourlès, D., Brown, R. T. and Chardon, D., 2005. Use of in situ-produced Be-10 in carbonate-rich environments: A first attempt, *Geochim. Cosmochim. Acta* **69** 1473–1478.

Braucher, R., Bourlès, D. L., Brown, E. T., Colin, F., Muller, J.-P., Braun, J.-J., Delaune, M., Edou Minko, A., Lescouet, C., Raisbeck, G. M. and Yiou, F., 2000. Application of in situ-produced cosmogenic ^{10}Be and ^{26}Al to the study of lateritic soil development in tropical forest: theory and examples from Cameroon and Gabon, *Chem. Geol.* **170** 95–111.

Braucher, R., Bourlès, D. L. and Colin, F., 1998a. Use of *in situ*-produced cosmogenic 10-Be for the study of Brazilian lateritic soil evolution, Annual Meeting of the Geological Society of America, GSA, Toronto.

Braucher, R., Bourlès, D. L., Colin, F., Brown, E. T. and Boulangé, B., 1998b. Brazilian laterite dynamics using in situ-produced ^{10}Be, *Earth Planet. Sci. Lett.* **163** 197–205.

Braucher, R., Colin, F., Brown, E. T., Bourles, D. L., Bamba, O., Raisbeck, G. M., Yiou, F. and Koud, J. M., 1998c. African laterite dynamics using *in situ*-produced Be-10, *Geochim. Cosmochim. Acta* **62** 1501–1507.

Braucher, R., Del Castillo, P., Siame, L., Hidy, A. J. and Bourlès, D., 2009. Determination of both exposure time and denudation rate from an in situ-produced ^{10}Be depth profile: A mathematical proof of uniqueness. Model sensitivity and applications to natural cases, *Quat. Geochronol.* **4** 56–67.

Briner, J. P., Miller, G. H., Davis, P. T. and Finkel, R. C., 2005. Cosmogenic exposure dating in arctic glacial landscapes: implications for the glacial history of northeastern Baffin Island, Arctic Canada, *Can. J. Earth Sci.* **42** 67–84.

Briner, J. P., Miller, G. H., Davis, P. T. and Finkel, R. C., 2006. Cosmogenic radionuclides from fiord landscapes support differential erosion by overriding ice sheets, *Geol. Soc. Am. Bull.* **118** 406–420.

Brook, E. J., Brown, E. T., Kurz, M. D., Ackert, R. P., Raisbeck, G. M. and Yiou, F., 1995a. Constraints on age, erosion and uplift of Neogene glacial deposits in the Transantarctic Mountains determined from in situ cosmogenic ^{10}Be and ^{26}Al, *Geology* **23** 1063–1066.

Brook, E. J., Kurz, M. D., Ackert, J. R. P., Denton, G. H., Brown, E. T., Raisbeck, G. M. and Yiou, F., 1993. Chronology of Taylor Glacier advances in Arena Valley, Antarctica, using *in situ* cosmogenic ^{3}He and ^{10}Be, *Quat. Res.* **39** 11–23.

Brook, E. J., Kurz, M. D., Ackert, R. P., Raisbeck, G. M. and Yiou, F., 1995b. Cosmogenic nuclide exposure ages and glacial history of late Quaternary Ross Sea drift in McMurdo Sound, Antarctica, *Earth Planet. Sci. Lett.* **131** 41–56.

Brown, E. T., Edmont, J. M., Raisbeck, G. M., Yiou, F., Kurz, M. D. and Brook, E. J., 1991. Examination of surface exposure ages of Antarctic moraines using in situ produced ^{10}Be and ^{26}Al, *Geochim. Cosmochim. Acta* **55** 2269–2283.

Brown, E. T., Stallard, R. F., Larsen, M. C., Raisbeck, G. M. and Yiou, F., 1995. Denudation rates determined from the accumulation of in situ produced ^{10}Be in the Luquillo Experimental forest, Puerto Rico, *Earth Planet. Sci. Lett.* **129** 193–202.

Brown, E. T., Trull, W. T., Jean-Baptiste, P., Raisbeck, G., Bourlès, D., Yiou, F. and Marty, B., 2000. Determination of cosmogenic production rates of ^{10}Be, ^{3}He, and ^{3}H in water, *Nucl. Inst. Meth. Phys. Res. B* **172** 873–883.

Brown, R. T., Brook, E. J., Raisbeck, G. M., Yiou, F. and Kurz, M. D., 1992. Effective attenuation length of cosmic rays producing ^{10}Be and ^{26}Al in quartz: implications for exposure dating, *Geophys. Res. Lett.* **19** 369–372.

Browne, J. C. and Berman, B. L., 1973. Neutron-capture cross-sections for ^{128}Te and ^{130}Te and the xenon anomaly in old tellurium ores, *Phys. Rev. C* **8** 154–154.

Bruno, L. A., Baur, H., Graf, T., Schlüchter, C., Signer, P. and Wieler, R., 1997. Dating of Sirius Group tillites in the Antarctic Dry valleys with cosmogenic ^{3}He and ^{21}Ne, *Earth Planet. Sci. Lett.* **147** 37–54.

Burbank, D. W., Leland, J., Fielding, E., Anderson, R. S., Brozovic, N., Reid, M. R. and Duncan, C., 1996. Bedrock incision, rock uplift and threshold hillslope in the northwestern Himalayas, *Nature* **379** 505–510.

Burke, B. C., Heimsath, A. M. and White, A. F., 2007. Coupling chemical weathering with soil production across soil-mantled landscapes, *Earth Surf. Process. Landforms* **32** 853–873.

Carizzo, D., González, G. and Dunai, T. J., 2008. Constricción neógena en la Cordillera de la Costa, norte de Chile: neotectónica y datación de superficies con ^{21}Ne cosmogénico, *Revista Geol. Chile* **35** 1–38.

Cerling, T. E. and Craig, H., 1994a. Cosmogenic ^{3}He production rates from 39°N to 46°N lattitude, western USA and France, *Geochim. Cosmochim. Acta* **58** 249–255.

Cerling, T. E. and Craig, H., 1994b. Geomorphology and in-situ cosmogenic isotopes, *Annu. Rev. Earth Planet. Sci.* **22** 273–317.

Cerling, T. E., Webb, R. H., Poreda, R. J., Rigby, A. D. and Melis, T. S., 1999. Cosmogenic ^{3}He ages and frequency of late Holocene debris flows from Prospect Canyon, Grand Canyon, USA, *Geomorphol.* **27** 93–111.

Chadwick, M. B., Obložinský, P., Herman, M., Greene, N. M., McKnight, R. D., Smith, D. L., Young, P. G., MacFarlane, R. E., Hale, G. M., Frankle, S. C., Kahler, A. C., Kawano, T., Little, R. C., Madland, D. G., Moller, P., Mosteller, R. D., Page, P. R., Talou, P., Trellue, H., White, M. C., Wilson, W. B., Arcilla, R., Dunford, C. L., Mughabghab, S. F., Pritychenko, B., Rochman, D., Sonzogni, A. A., Lubitz, C. R., Trumbull, T. H., Weinman, J. P., Brown, D. A., Cullen, D. E., Heinrichs, D. P., McNabb, D. P., Derrien, H., Dunn, M. E., Larson, N. M., Leal, L. C., Carlson, A. D., Block, R. C., Briggs, J. B., Cheng, E. T., Huria, H. C., Zerkle, M. L., Kozier, K. S., Courcelle, A., Pronyaev, V. and van der Marck, S. C., 2006. ENDF/B-VII.0: Next Generation Evaluated Nuclear Data Library for Nuclear Science and Technology, *Nucl. Data Sheets* **107** 2931–3060.

Charalambus, S., 1971. Nuclear transmutation by negative stopped muons and the activity induced by the cosmic-ray muons, *Nucl. Phys.* **A166** 145–161.

Chemeleff, J., von Blanckenburg, F., Kossert, K. and Jakob, D., 2009. Determination of the [10]Be half-life by Multi Collector ICP-Mass Spectrometry and Liquid Scintillation Counting, *Geochim. Cosmochim. Acta.* **73** A221.

Clark, D. H., Bierman, P. R. and Larsen, P., 1995. Inproving *in situ* cosmogenic chronometers, *Quat. Res.* **44** 367–377.

Clarke, W. B., Beg, M. A. and Craig, H., 1969. Excess [3]He in the sea: evidence for terrestrial primordial helium, *Earth Planet. Sci. Lett.* **6** 213–220.

Clem, J. M. and Dorman, L. I., 2000. Neutron monitor response functions, *Space Sci. Rev.* **93** 335–359.

Codilean, A. T., 2006. Calculation of the cosmogenic nuclide production topographic shielding scaling factor for large areas using DEMs, *Earth Surf. Process. Landforms* **31** 785–794.

Codilean, A. T., Bishop, P., Stuart, F. M., Hoey, T. B., Fabel, D. and Freeman, S. P. H. T., 2008. Single-grain cosmogenic [21]Ne concentrations in fluvial sediments reveal spatially variable erosion rates, *Geology* **36** 159–162.

Conversi, M. and Rothwell, P., 1954. Angular distribution in cosmic ray stars at 3500 meters, *Nuovo Cimiento* **12** 191.

Craig, H. and Poreda, R. J., 1986. Cosmogenic [3]He in terrestrial rocks: the summit lavas of Maui, *Proc. Natl. Acad. Sci.* **83** 1970–1974.

Daeron, M., Benedetti, L., Tapponnier, P., Sursock, A. and Finkel, R. C., 2004. Constraints on the post-25-ka slip rate of the Yammouneh fault (Lebanon) using in situ cosmogenic Cl-36 dating of offset limestone-clast fans, *Earth Planet. Sci. Lett.* **227** 105–119.

Dalrymple, G. B. and Lanphere, M. A., 1969 *Potassium-Argon Dating*, Freeman, San Francisco, 258 pp.

Damon, P. E. and Jirikowic, J. L., 1992. The sun as a low-frequency harmonic oscillator, *Radiocarbon* **34** 199–205.

Davis, P. T., Bierman, P. R., Marsella, K. A., Caffee, M. W. and Southon, J. R., 1999. Cosmogenic analysis of glacial terrains in the eastern Canadian Arctic: a test for inherited nuclides and the effectiveness of glacial erosion, *Ann. Glaciol.* **28** 181–188.

Davis, R. and Schaeffer, O. A., 1955. Chlorine-36 in nature, *Ann. NY Acad. Sci.* **62** 105–122.

de Laeter, J. R., 1998. Mass spectrometry and geochronology, *Mass Spec. Rev.* **17** 97–125.

Deer, W. A., Howie, R. A. and Zussman, J., 1992. *The Rock Forming Minerals*, New York: John Wiley & Sons, Inc., 696 pp.

Desilets, D. and Zreda, M., 2003. Spatial and temporal distribution of secondary cosmic-ray nucleon intensities and applications to in situ cosmogenic dating, *Earth Planet. Sci. Lett.* **206** 21–42.

Desilets, D., Zreda, M., Almasi, P. F. and Elmore, D., 2006a. Determination of cosmogenic [36]Cl in rocks by isotope dilution: innovations, validation and error propagation, *Chem. Geol.* **233** 185–195.

Desilets, D., Zreda, M. and Lifton, N. A., 2001. Comment on "Scaling factors for production rates of in situ produced cosmogenic nuclides: a critical reevaluation" by Tibor J. Dunai, *Earth Planet. Sci. Lett.* **188** 283–287.

Desilets, D., Zreda, M. and Prabu, T., 2006b. Extended scaling factors for in situ cosmogenic nuclides: New measurements at low latitude, *Earth Planet. Sci. Lett.* **246** 265–276.

Desilets, D., Zreda, M. and Terré, T., 2007. Scientist water equivalent measured with cosmic rays at 2006 AGU Fall Meeting, *Eos Trans. AGU* **88** 521–536.

Diehl, R., Halloin, H., Kretschmer, K., Lichti, G. G., Schönfelder, V., Strong, A. W., von Kienlin, A., Wang, W., Jean, P., Knödlseder, J., Roques, J. P., Weidenspointner, G., Schanne, S., Hartmann, D. H., Winkler, C. and Wunderer, C., 2006. Radioactive Al-26 from massive stars in the Galaxy, *Nature* **439** 45–47.

Dixit, K. R., 1955. The statistics of 29000 stars observed in nuclear emulsions in Kenya, *Z. Naturforschung* **10** 339–341.

Dorman, L. I., Valdés-Galicia, J. F. and Dorman, I. V., 1999. Numerical simulation and analytical description of solar neutron transport in the Earth's atmosphere, *J. Geophys. Res.* **104** 22417–22426.

Dorman, L. I., Villoresi, G., Iucci, N., Parisi, M., Tyasto, M. I., Danilova, O. A. and Ptitsyna, N. G., 2000. Cosmic ray survey to Antarctica and coupling functions for neutron component near solar minimum (1996–1997) 3. Geomagnetic effects and coupling functions, *J. Geophys. Res.* **105** 21047–21056.

Dugan, B., Lifton, N. and Jull, A. J. T., 2008. New production rate estimates for in situ cosmogenic ^{14}C, *Geochim. Cosmochim. Acta* **72** A231.

Duhnforth, M., Densmore, A. L., Ivy-Ochs, S., Allen, P. A. and Kubik, P. W., 2007. Timing and patterns of debris flow deposition on Shepherd and Symmes creek fans, Owens Valley, California, deduced from cosmogenic Be-10, *J. Geophys. Res. Earth Surf.* **112** F03S15.

Dunai, T. J., 2000. Scaling factors for production rates of in-situ produced cosmogenic nuclides: a critical reevaluation, *Earth Planet. Sci. Lett.* **176** 157–169.

Dunai, T. J., 2001a. Influence of secular variation of the geomagnetic field on production rates of in-situ produced cosmogenic nuclides, *Earth Planet. Sci. Lett.* **193** 197–212.

Dunai, T. J., 2001b. Reply to comment on "Scaling factors for production rates of in situ produced cosmogenic nuclides: a critical reevaluation" by Darin Desilets, Marek Zreda and Nathaniel Lifton, *Earth Planet. Sci. Lett.* **188** 289–298.

Dunai, T. J., González López, G. A. and Juez-Larré, J., 2005. Oligocene-Miocene age of aridity in the Atacama Desert revealed by exposure dating of erosion-sensitive landforms, *Geology* **33** 311–324.

Dunai, T. J. and Porcelli, D., 2002. Storage and transport of noble gases in the subcontinental lithosphere. In: D. Porcelli, C. Ballentine and R. Wieler, (Eds), *Noble Gases in Cosmochemistry and Geochemistry*, Reviews in Mineralogy and Geochemistry 47, Washington: The Mineralogical Society of America, pp. 371–410.

Dunai, T. J. and Roselieb, K., 1996. Sorption and diffusion of helium in garnet: implications for volatile tracing and dating, *Earth Planet. Sci. Lett.* **139** 411–421.

Dunai, T. J. and Stuart, F. M., 2009. Reporting of cosmogenic nuclide data for exposure age and erosion rate determinations, *Quat. Geochronol.* doi:10.1016/j.quageo.2009.1004.1003.

Dunai, T. J., Stuart, F. M., Pik, R., Burnard, P. G. and Gayer, E., 2007. Production of ^{3}He in crustal rocks by cosmogenic thermal neutrons, *Earth Planet. Sci. Lett.* **258** 228–236.

Dunai, T. J. and Wijbrans, J. R., 2000. Long-term cosmogenic ^{3}He production rates (152 ka–1.35 Ma) from ^{40}Ar/^{39}Ar dated basalt flows at 29°N latitude, *Earth Planet. Sci. Lett.* **176** 147–156.

Dunne, J., Elmore, D. and Muzikar, P., 1999. Scaling factors for the rates of production of cosmogenic nuclides for geometric shielding and attenuation at depth on sloped surfaces, *Geomorphol.* **27** 3–11.

Eberhardt, P., Eugster, O. and Marti, K., 1965. A redetermination of the isotopic composition of atmospheric neon, Z. *Naturforschung* **20a** 623–624.

Eidelman, S., Hayes, K. G., Olive, K. A., Aguilar-Benitez, M., Amsler, C., Asner, D., Babu, K. S., Barnett, R. M., Beringer, J., Burchat, P. R., Carone, C. D., Caso, C., Conforto, G., Dahl, O., D'Ambrosio, G., Doser, M., Feng, J. L., Gherghetta, T., Gibbons, L., Goodman, M., Grab, C., Groom, D. E., Gurtu, A., Hagiwara, K., Hernandez-Rey, J. J., Hikasa, K., Honscheid, K., Jawahery, H., Kolda, C., Kwon, Y., Mangano, M. L., Manohar, A. V., March-Russell, J. and Masoni, A., 2004. Review of particle physics, *Phys. Lett. B* **592** 1–1109.

Elsasser, W., Ney, E. P. and Winckler, J. R., 1956. Cosmic-ray intensity and geomagnetism, *Nature* **178** 1226–1227.

Eugster, O., 1988. Cosmic-ray production-rates for He-3, Ne-21, Ar-38, Kr-83, and Xe-126 in chondrites based on 81Kr exposure ages, *Geochim. Cosmochim. Acta* **52** 1649–1662.

Evans, J. M., 2002. *Calibration of the Production Rates of Cosmogenic ^{36}Cl from Potassium*, Ph.D. dissertation, Australian National University.

Evans, J. M., Stone, J. O. H., Fifield, L. K. and Cresswell, R. G., 1997. Cosmogenic chlorine-36 production in K-feldspar, *Nucl. Instr. Meth. Phys. Res. B* **123** 334–340.

Evenstar, L. A., Hartley, A. J., Stuart, F. M., Mather, A. E., Rice, C. M. and Chong, G., 2009. Multiphase development of the Atacama Planation Surface recorded by cosmogenic He-3 exposure ages: Implications for uplift and Cenozoic climate change in western South America, *Geology* **37** 27–30.

Fabel, D., Stroeven, A. P., Harbor, J., Kleman, J., Elmore, D. and Fink, D., 2002. Landscape preservation under Fennoscandian ice sheets determined from in situ produced Be-10 and Al-26, *Earth Planet. Sci. Lett.* **201** 397–406.

Farber, D. L., Meriaux, A. S. and Finkel, R. C., 2008. Attenuation length for fast nucleon production of ^{10}Be derived from near-surface production profiles, *Earth Planet. Sci. Lett.* **274** 295–300.

Farley, K. A., 2007. He diffusion systematics in minerals: Evidence from synthetic monazite and zircon structure phosphates, *Geochim. Cosmochim. Acta* **71** 4015–4052.

Farley, K. A., Libarkin, J., Mukhopadhyay, S. and Amidon, W., 2006. Cosmogenic and nucleogenic ^{3}He in apatite, titanite, and zircon, *Earth Planet. Sci. Lett.* **248** 451–461.

Farley, K. A., Wolf, R. A. and Silver, L. T., 1996. The effects of long alpha-stopping distances on (U-Th)/He ages, *Geochim. Cosmochim. Acta* **60** 4223–4229.

Faure, G. and Mensing, T. M., 2004. *Isotopes: Principles and Applications*, Chichester: John Wiley & Sons, Ltd, 928 pp.

Filges, D., Goldenbaum, F., Enke, M., Galin, J., Herbach, C. M., Hilscher, D., Jahnke, U., Letourneau, A., Lott, B., Neef, R. D., Nunighoff, K., Paul, N., Peghaire, A., Pienkowski, L., Schaal, H., Schroder, U., Sterzenbach, G., Tietze, A., Tishchenko, V., Toke, J. and Wohlmuther, M., 2001. Spallation neutron production and the current intra-nuclear cascade and transport codes, *Eur. Phys. J. A* **11** 467–490.

Fodor, L., Bada, G., Csillag, G., Horváth, E., Ruszkiczay-Rüdiger, Z., Palotás, K., Síkhegyi, F., Timár, G., Cloetingh, S. and Horváth, F., 2005. An outline of neotectonics and morphotectonics of the western and central Pannonian Basin, *Tectonophys.* **410** 15–41.

Fortier, S. M. and Giletti, B. J., 1989. An empirical model for predicting diffusion coefficients in silicate minerals, *Science* **245** 1481–1484.

Frankel, K. L., Brantley, K. S., Dolan, J. F., Finkel, R. C., Klinger, R. E., Knott, J. R., Machette, M. N., Owen, L. A., Phillips, F. M., Slate, J. L. and Wernicke, B. P., 2007. Cosmogenic Be-10 and Cl-36 geochronology of offset alluvial fans along the northern Death Valley fault zone: Implications for transient strain in the eastern California shear zone, *J. Geophys. Res. Earth Surf.* **112** B06407.

Gao, X., Gao, B., Shen, G. and Granger, D., 2009. Age of Zhoukoudian determined with ^{26}Al/^{10}Be burial dating, *Nature* **458** 198–200.

Gayer, E., Mukhopadhyay, S. and Meade, B. J., 2008. Spatial variability of erosion rates inferred from the frequency distribution of cosmogenic He-3 in olivines from Hawaiian river sediments, *Earth Planet. Sci. Lett.* **266** 303–315.

Gilbert, G. K., 1877. *Report on the Geology of the Henry Mountains*, Washington DC: Government Printing Office, 160 pp.

Gladkis, L. G., Fifield, L. K., Morton, C., Barrows, T. T. and Tims, S. G., 2007. Manganese-53: Development of the AMS technique for exposure-age dating applications, *Nucl. Instr. Meth. Phys. Res. B* **259** 236–240.

Goethals, M., Hetzel, R., Niedermann, S., Wittmann, H., Fenton, C. R., Christl, M., Kubik, P. and von Blanckenburg, F., 2009a. An improved experimental determination of cosmogenic ^{10}Be/^{21}Ne and ^{26}Al/^{21}Ne production ratios in quartz, *Earth Planet. Sci. Lett.* doi:10.1016/j.epsl.2009.1004.1027.

Goethals, M., Niedermann, S., Hetzel, R. and Fenton, C. R., 2009b. Determining the impact of faulting on the rate of erosion in a low-relief landscape: A case study using in situ produced Ne-21 on active normal faults in the Bishop Tuff, California, *Geomorphol.* **103** 401–413.

Goldhagen, P., Reginatto, M., Kniss, T., Wilson, J. W., Singleterry, R. C., Jones, I. W. and Van Steveninck, W., 2002. Measurement of the energy spectrum of cosmic-ray induced neutrons aboard an ER-2 high-altitude airplane, *Nucl. Instr. Methods Phys. Res. A* **476** 42–51.

Goldstein, S. L., Deines, P., Oelkers, E. H., Rudnick, R. L. and Walter, L. M., 2003. Standards for publication of isotopic ratio and chemical data in Chemical Geology, *Chem. Geol.* **202** 1–4.

González, G., Dunai, T. J., Carrizo, D. and Allmendinger, R., 2006. Young displacements on the Atacama Fault System, northern Chile from field observations and cosmogenic Ne-21 concentrations, *Tectonics* **25** TC3006.

Gordon, M. S., Goldhagen, P., Rodbell, K. P., Zabel, T. H., Tang, H. H. K., Clem, J. M. and Bailey, P., 2004. Measurement of the flux and energy spectrum of cosmic-ray induced neutrons on the ground, *IEEE Trans. Nucl. Sci.* **51** 3427–3434.

Gosse, J. C. and Phillips, F. M., 2001. Terrestrial in situ cosmogenic nuclides: theory and application, *Quat. Sci. Rev.* **20** 1475–1560.

Gosse, J. C., Reedy, R. C., Harrington, C. D. and Poths, J., 1996. Overview of the workshop on secular variations in production rates of cosmogenic nuclides on Earth, *Radiocarbon* **38** 135–147.

Gosse, J. C. and Stone, J. O., 2001. Terrestrial cosmogenic nuclide methods passing milestones toward paleo-altimetry, *EOS, Trans. Am. Geophys. Union* **82** 82.

Gould, B. A., 1855. On Peirce's criterion for the rejection of doubtful observations, with tables for facilitating its application, *Astronom. J.* **4** 81–87.

Graf, A. A., Strasky, S., Ivy-Ochs, S., Akcar, N., Kubik, P., Burkhard, M. and Schlüchter, C., 2007. First results of cosmogenic dated pre-Last Glaciation erratics from the Montoz area, Jura Mountains, Switzerland, *Quat. Int.* **164–165** 43–52.

Graham, D. W., 2002. Noble gas isotopic geochemistry of mid-ocean ridge and ocean island basalts: Characterization of mantle source reservoirs, *Rev. Mineral. Geochem.* **47** 247–318.

Granger, D., 2006. A review of burial dating methods using ^{26}Al and ^{10}Be. In: L. Siame, D. L. Bourles and E. T. Brown, (Eds) *In Situ-Produced Cosmogenic Nuclides and Quantification of Geological Surfaces* Special paper 415, Boulder: The Geological Society of America, pp. 1–16.

Granger, D., Fabel, D. and Palmer, A. N., 2001. Pliocene-Pleistocene incision of the Green River, Kentucky, determined from radioactive decay of cosmogenic Al-26 and Be-10 in Mammoth Cave sediments, *Geol. Soc. Am. Bull.* **113** 825–836.

Granger, D. and Riebe, C. S., 2007. Cosmogenic nuclides in weathering and erosion, *Treatise on Geochemistry*, Amsterdam: Elsevier, Chapter 5.19, pp. 11–43.

Granger, D. E., Kirchner, J. W. and Finkel, R., 1996. Spatially averaged long-term erosion rates measured from in-situ produced cosmogenic nuclides in alluvial sediment, *J. Geol.* **104** 249–257.

Granger, D. E., Kirchner, J. W. and Finkel, R. C., 1997. Quaternary downcutting rate of the New River, Virginia, measured from differential decay of cosmogenic ^{26}Al and ^{10}Be in cave-deposited alluvium, *Geology* **25** 107–110.

Granger, D. E. and Muzikar, P. F., 2001. Dating sediment burial with in-situ produced cosmogenic nuclides: theory, techniques, and limitations, *Earth Planet. Sci. Lett.* **1888** 269–281.

Granger, D. E. and Smith, A. L., 1998. Early Laurentide glaciation and creation of the Ohio river dated by radioactive decay of cosmogenic Al-26 and Be-10 in proglacial sediments, Annual Meeting of the Geological Society of America, GSA, Toronto.

Granger, D. E. and Smith, A. L., 2000. Dating buried sediments using radioactive decay and muogenic production of ^{26}Al and ^{10}Be, *Nucl. Instr. Meth. Phys. Res. B* **172** 822–826.

Groom, D. E., Mokhov, N. V. and Striganov, S. I., 2001. Muon stopping-power and range tables, *Atom. Data Nucl. Data Tables* **78** 183–356.

Grosse, A., 1934. An unknown radioactivity, *J. Am. Chem. Soc.* **56** 1922–1924.

Guedes, S., Jonckheere, R., Iunes, P. J. and Hadler, J. C., 2007. Projected-length distributions of fission-fragment tracks from U and Th thin film sources in muscovite, *Nucl. Instr. Meth. Phys. Res. B* **266** 786–790.

Guyodo, Y. and Valet, J.-P., 1999. Global changes in intensity of the Earth's magnetic field during the past 800 kyr, *Nature* **399** 249–252.

Haeuselmann, P., Granger, D. E., Jeannin, P. Y. and Lauritzen, S. E., 2007. Abrupt glacial valley incision at 0.8 Ma dated from cave deposits in Switzerland, *Geology* **35** 143–146.

Hancock, G. S., Anderson, R. S., Chadwick, O. A. and Finkel, R. C., 1999. Dating fluvial terraces with ^{10}Be and ^{26}Al profiles: application to the Wind River, Wyoming, *Geomorphol.* **27** 41–60.

Handwerger, D. A., Cerling, T. E. and Bruhn, R. L., 1999. Cosmogenic ^{14}C in carbonate rocks, *Geomorphol.* **27** 13–24.

Harbor, J., Stroeven, A. P., Fabel, D., Clarhall, A., Kleman, J., Li, Y. K., Elmore, D. and Fink, D., 2006. Cosmogenic nuclide evidence for minimal erosion across two subglacial sliding boundaries of the late glacial Fennoscandian ice sheet, *Geomorphol.* **75** 90–99.

Hatton, C. J., 1971. The neutron monitor. In: J. G. Wilson and S. A. Wouthuysen, (Eds), *Progress in Elementary Particle and Cosmic Ray Physics* 10, Amsterdam: North Holland, pp. 3–100.

Hatton, C. J. and Carmichael, H., 1964. Experimental investigation of NM-64 neutron monitor, *Can. J. Phys.* **42** 2443–2472.

Heidbreder, E., Pinkau, K., Reppin, C. and Schönfelder, V., 1971. Measurements of the distribution in energy and angle of high-energy neutrons in the lower atmosphere, *J. Geophys. Res.* **76** 2905–2916.

Heimsath, A. M., Chappell, J., Spooner, N. A. and Questiaux, D. G., 2002. Creeping soil, *Geology* **30** 111–114.

Heimsath, A. M., Dietrich, W. E., Nishiizumi, K. and Finkel, R. C., 1999. Cosmogenic nuclides, topography, and the spatial variation of soil depth, *Geomorphol.* **27** 151–172.

Heimsath, A. M., Dietrich, W. E., Nishiizumi, K. and Finkel, R. C., 2001. Stochastic processes of soil production and transport: Erosion rates, topographic variation and cosmogenic nuclides in the Oregon Coast Range, *Earth Surf. Process. Landforms* **26** 531–552.

Heimsath, A. M., Dietrich, W. E., Nishiizumi, K. and Finkel, R. C., 1997. The soil production function and landscape equilibrium, *Nature* **388** 358–361.

Heimsath, A. M., Furbish, D.J and Dietrich, W. E., 2005. The illusion of diffusion: Field evidence for depth-dependent sediment transport, *Geology* **33** 949–952.

Hein, A. S., Hulton, N. R. J., Dunai, T. J., Schnabel, C., Kaplan, M. R., Xu, S. and Naylor, M., 2009. Middle Pleistocene glaciation in Patagonia dated by cosmogenic-nuclide measurements on outwash gravels, *Earth Planet. Sci. Lett.* doi:10.1016/j.epsl.2009.06.026.

Heisinger, B., Lal, D., Jull, A. J. T., Kubik, P., Ivy-Ochs, S., Knie, K. and Nolte, E., 2002a. Production of selected cosmogenic radionuclides by muons: 2. Capture of negative muons, *Earth Planet. Sci. Lett.* **200** 357–369.

Heisinger, B., Lal, D., Jull, A. J. T., Kubik, P., Ivy-Ochs, S., Neumaier, S., Knie, K., Lazarev, V. and Nolte, E., 2002b. Production of selected cosmogenic radionuclides by muons 1. Fast muons, *Earth Planet. Sci. Lett.* **200** 345–355.

Hellborg, R. and Skog, G., 2008. Accelerator mass spectrometry, *Mass Spec. Rev.* **27** 398–427.

Herber, L. J., 1969. Separation of feldspar from quartz by flotation, *Am. Mineral.* **54** 1212–1214.

Hermanns, R. L., Niedermann, S., Garcia, A. V., Gomes, J. S. and Strecker, M. R., 2001. Neotectonics and catastrophic failure of mountain fronts in the southern intra-Andean Puna Plateau, Argentia, *Geology* **29** 619–622.

Hetzel, R., Niedermann, S., Ivy-Ochs, S., Kubik, P. W., Tao, M. X. and Gao, B., 2002a. Ne-21 versus Be-10 and Al-26 exposure ages of fluvial terraces: the influence of crustal Ne in quartz, *Earth Planet. Sci. Lett.* **201** 575–591.

Hetzel, R., Niedermann, S., Tao, M., Kubik, P. W., Ivy-Ochs, S., Bao, B. and Strecker, M. R., 2002b. Low slip rates and long-term preservation of geomorphic features in Central Asia, *Nature* **417** 428–432.

Hewawasam, T., von Blanckenburg, F. and Schaller, M., 2003. Increase of human over natural erosion rates in tropical highlands constrained by cosmogenic isotopes, *Geology* **31** 597–600.

Hilton, D. R., Hammerschmidt, K., Teufel, S. and Friedrichsen, H., 1993. Helium isotope characteristics of Andean geothermal fluids and lavas, *Earth Planet. Sci. Lett.* **120** 265–282.

Hiyagon, H., 1994. Retention of solar helium and neon in IDPs in deep sea sediment, *Science* **263** 1257–1259.

Hofmann, H. J., Beer, J., Bonani, G., von Gunten, H. R., Raman, S., Suter, M., Walker, R. L., Wölfli, W. and Zimmermann, D., 1987. ^{10}Be: half life and AMS-standards, *Nucl. Instr. Meth. Phys. Res.* **B29** 32–36.

Hohenberg, C. M., Marti, K., Podosek, F. A., Reedy, R. C. and Shirck, J. R., 1978. Comparisons between observed and predicted cosmogenic noble gases in lunar samples, *Proceedings of the 9th Lunar and Planetary Science Conference*, LPI, Houston, pp. 2311–2344.

Holden, N. E., 1990. Total half-lives for selected nuclides, *Pure Appl. Chem.* **62** 941–958.

Honda, M. and Imamura, M., 1971. Half-life of ^{53}Mn, *Phys. Rev. C* **4** 1182–1188.

Humphreys, G. S. and Wilkinson, M. T., 2007. The soil production function: A brief history and its rediscovery, *Geoderma* **139** 73–78.

Ivy-Ochs, S., Kober, F., Alfimov, V., Kubik, P. and Synal, H. A., 2007. Cosmogenic ^{10}Be, ^{21}Ne and ^{36}Cl in sanidine and quartz from Chilean ignimbrites, *Nucl. Instr. Meth. Phys. Res. B* **259** 588–594.

Jamieson, S. S. R., Hulton, N. R. J. and Hagdorn, M., 2008. Modelling landscape evolution under ice sheets, *Geomorphol.* **97** 91–108.

Jull, A. J. T., Barker, D. L. and Donahue, D. J., 1987. On the ^{14}C content in radioactive ores, *Chem. Geol.* **66** 35–40.

Jull, A. J. T., Lifton, N., Phillips, W. M. and Quade, J., 1994. Studies of the production rate of cosmic-ray produced ^{14}C in rock surfaces, *Nucl. Instr. Meth. Phys. Res. B* **92** 308–310.

Jull, A. J. T., Wilson, A. E., Donahue, D. J., Toolin, L. J. and Burr, G. S., 1992. Measurement of cosmogenic ^{14}C produced by spallation in high-altitude rocks, *Radiocarbon* **34** 737–744.

Kaste, J. M., Heimsath, A. M. and Bostick, B. C., 2007. Short-term soil mixing quantified with fallout radionuclides, *Geology* **35** 243–246.

Kennedy, B. M., Hiyagon, H. and Reynolds, J. H., 1990. Crustal neon: a striking uniformity, *Earth Planet. Sci. Lett.* **98** 277–286.

Kim, K. J., Lal, D., Englert, P. A. J. and Southon, J., 2007. *In situ* C-14 depth profile of subsurface vein quartz samples from Macraes Flat New Zealand, *Nucl. Instr. Meth. Phys. Res. B* **259** 632–636.

Klein, J., Giegengack, R., Middleton, R., Sharma, P., Underwood, J. R. and Weeks, W. A., 1986. Revealing histories of exposure using in-situ produced ^{26}Al and ^{10}Be in Libyan desert glass, *Radiocarbon* **28** 547–555.

Knight, J., 2008. The environmental significance of ventifacs: A critical review, *Earth Sci. Rev.* **86** 89–105.

Knudsen, M. F., Riisager, P., Donadini, F., Snowball, I., Muscheler, R., Korhonen, K. and Pesonen, L. J., 2008. Variations in the geomagnetic dipole moment during the Holocene and the past 50 kyr, *Earth Planet. Sci. Lett.* **272** 319–329.

Kober, F., Ivy-Ochs, S., Leya, I., Baur, H., Magna, T., Wieler, R. and Kubik, P. W., 2005. In situ cosmogenic ^{10}Be and ^{21}Ne in sanidine and *in situ* cosmogenic ^{3}He in Fe-Ti-oxide minerals, *Earth Planet. Sci. Lett.* **236** 404–418.

Kober, F., Ivy-Ochs, S., Schlunegger, F., Baur, H., Kubik, P. W. and Wieler, R., 2007. Denudation rates and a topography-driven rainfall threshold in northern Chile: Multiple cosmogenic nuclide data and sediment yield budgets, *Geomorphol.* **83** 97–120.

Kohl, C. P. and Nishiizumi, K., 1992. Chemical isolation of quartz for measurement of in-situ-produced cosmogenic nuclides, *Geochim. Cosmochim. Acta* **56** 3583–3587.

Kong, P., Granger, D., Wu, F., Caffee, M. W., Wang, Y., Zhao, X. and Zheng, Y., 2009. Cosmogenic nuclide burial ages and provenance of the Xigeda paleo-lake: Implications for evolution of the Middle Yangtze River, *Earth Planet. Sci. Lett.* **278** 131–141.

Koppers, A. A. P., Staudigel, H. and Wijbrans, J. R., 2000. Dating crystalline groundmass separates of altered Cretaceous seamount basalts by the ^{40}Ar/^{39}Ar incremental heating technique, *Chem. Geol.* **166** 139–158.

Korschinek, G., Bergmaier, A., Dillmann, I., Faestermann, T., Gerstmann, U., Knie, K., von Gostomski, C. L., Maiti, M., Poutivtsev, M., Remmert, A., Rugel, G. and Wallner, A., 2009. Determination of the ^{10}Be half-life by HI-IRD and liquid scintillation counting, *Geochim. Cosmochim. Acta* **73** A685.

Korte, M. and Constable, C., 2005. Continuous geomagnetic field models for the past 7 millennia: 2. CALS7K, *Geochem. Geophys. Geosyst.* **6** 1–18.

Kowatari, M., Nagaoka, K., Satoh, S., Ohta, Y., Abukawa, J., Tachimori, S. and Nakamura, T., 2005. Evaluation of the altitude variation of the cosmic-ray induced environmental neutrons in the Mt. Fuji area, *J. Nucl. Sci. Technol.* **42** 495–502.

Kubik, P. W., Ivy-Ochs, S., Masarik, J., Frank, M. and Schlüchter, C., 1998. ^{10}Be and ^{26}Al production rates deduced from an instantaneous event within the dendro-calibration curve, the landslide of Köfels, Ötz Valley, Austria, *Earth Planet. Sci. Lett.* **161** 231–241.

Kurz, M. D., 1986a. Cosmogenic helium in terrestrial igneous rock, *Nature* **320** 435–439.

Kurz, M. D., 1986b. *In situ* production of terrestrial cosmogenic helium and some applications to geochronology, *Geochim. Cosmochim. Acta* **50** 2855–2862.

Kurz, M. D., Colodner, D., Trull, T. W., Moore, R. B. and O'Brien, K., 1990. Cosmic ray exposure dating with in situ produced cosmogenic ^{3}He: results from young Hawaiian lava flows, *Earth Planet. Sci. Lett.* **97** 177–189.

Kutschera, W., 2005. Progress in isotope analysis at ultra-trace level by AMS, *Int. J. Mass Spec.* **242** 145–160.

Kutschera, W., Ahmad, I. and Paul, M., 1992. Half-life determinations of ^{41}Ca and some other radioisotopes, *Radiocarbon* **34** 436–446.

Lahiri, S., Nayak, D. and Korschinek, G., 2006. Separation of no-carrier-added ^{52}Mn from bulk chromium: A simulation study for accelerator mass spectrometry measurement of ^{53}Mn, *Anal. Chem.* **78** 7517–7521.

Lal, D., 1958. *Investigations of Nuclear Interactions Produced by Cosmic Rays*, PhD Thesis, Bombay University.

Lal, D., 1987. Production of ^{3}He in terrestial rocks, *Chem. Geol.* **66** 89–98.

Lal, D., 1988. *In situ*-produced cosmogenic isotopes in terrestrial rocks, *Ann. Rev. Earth Planet. Sci.* **16** 355–388.

Lal, D., 1991. Cosmic ray labeling of erosion surfaces: in situ nuclide production rates and erosion models, *Earth Planet. Sci. Lett.* **104** 424–439.

Lal, D. and Arnold, J. R., 1985. Tracing quartz through the environment, *Proc. Indian Acad. Sci. Earth Planet. Sci.* **94** 1–5.

Lal, D. and Jull, A. J. T., 1994. Studies of cosmogenic in-situ (CO)-C-14 and (CO_2)-C-14 produced in terrestrial and extraterrestrial samples – experimental procedures and applications, *Nucl. Instr. Meth. Phys. Res. B* **94** 291–296.

Lal, D., Malhotra, P. K. and Peters, B., 1958. On the production of radioisotopes in the atmosphere by cosmic radiation and their application to meteorology, *J. Atmos. Terrest. Phys.* **12** 306–328.

Lal, D. and Peters, B., 1962. Cosmic ray produced isotopes and their application to problems in geophysics. In: J. G. Wilson and S. A. Wouthuysen, (Eds), *Progress in Elementary Particle and Cosmic Ray Physics* **6**, Amsterdam: North Holland Publishing Company, pp. 77–243.

Lal, D. and Peters, B., 1967. Cosmic ray produced radioactivity on Earth. In: S. Flugg, (Ed), *Handbook of Physics* **46/2**, Berlin: Springer, pp. 551–612.

Lancaster, N., Kocurek, G., Singhvi, A., Pandey, V., Deynoux, M., Ghienne, J. F. and Lo, K., 2002. Late Pleistocene and Holocene dune activity and wind regimes in the western Sahara Desert of Mauritania, *Geology* **30** 991–994.

Lebatard, A. E., Bourles, D. L., Duringer, P., Jolivet, M., Braucher, R., Carcaillet, J., Schuster, M., Arnaud, N., Monie, P., Lihoreau, F., Likius, A., Mackaye, H. T., Vignaud, P. and Brunet, M., 2008. Cosmogenic nuclide dating of *Sahelanthropus tchadensis* and *Australopithecus bahrelghazali:* Mio-Pliocene hominids from Chad, *Proc. Natl. Acad. Sci. USA* **105** 3226–3231.

Lederer, C. M., Shirley, V. S., Browne, W., Diairiki, J. M., Doebler, R. E., Shihab-Eldin, A. A., Jardine, L. J., Tuli, J. K. and Buyrn, A. B., 1978. *Table of Isotopes*, New York: John Wiley & Sons, Inc., 690 pp.

Leya, I., Busemann, H., Baur, H., Wieler, R., Gloris, M., Neurmann, S., Michel, R., Sudbrock, F. and Herpers, U., 1998. Cross sections for the proton-induced production of He and Ne isotopes from magnesium, aluminium, and silicon, *Nucl. Instr. Meth. Phys. Res. B* **145** 449–458.

Leya, I., Lange, H. J., Neumann, S., Wieler, R. and Michel, R., 2000. The production of cosmogenic nuclides in stony meteorites by galactic cosmic ray particles, *Meteoritics Planet. Sci.* **35** 259–286.

Li, Y. K., Fabel, D., Stroeven, A. P. and Habor, J., 2008. Unraveling complex exposure-burial histories of bedrock surfaces under ice sheets by integrating cosmogenic nuclide concentrations with climate proxy records, *Geomorphol.* **99** 139–149.

Libby, W., 1946. Atmospheric helium three and radiocarbon from cosmic radiation, *Phys. Rev.* **69** 671–672.

Libby, W., Anderson, E. C. and Arnold, J. R., 1949. Age determination by radiocarbon content – world wide essay of natural radiocarbon, *Science* **109** 227–228.

Licciardi, J. M., Denoncourt, C. L. and Finkel, R. C., 2008. Cosmogenic ^{36}Cl production rates from Ca spallation in Iceland, *Earth Planet. Sci. Lett.* **267** 365–377.

Licciardi, J. M., Kurz, M. D., Clark, P. U. and Brook, E. J., 1999. Calibration of cosmogenic ^{3}He production rates from Holocene lava flows in Oregon, USA, and effects of the Earth's magnetic field, *Earth Planet. Sci. Lett.* **172** 261–271.

Lifton, N., 2008. In situ cosmogenic C-14 from surfaces at secular equilibrium, *Geochim. Cosmochim. Acta* **72** A552.

Lifton, N., Bieber, J. W., Clem, J. M., Duldig, M. L., Evenson, P., Humble, J. E. and Pyle, R., 2005. Addressing solar modulation and long-term uncertainties in scaling secondary cosmic rays for in situ cosmogenic nuclide applications, *Earth Planet. Sci. Lett.* **239** 140–161.

Lifton, N., Jull, A. J. T. and Quade, J., 2001. A new extraction technique and production rate estimate for in situ cosmogenic ^{14}C in quartz, *Geochim. Cosmochim. Acta* **65** 1953–1969.

Lifton, N. A., Smart, D. F. and Shea, M. A., 2008. Scaling time-integrated in situ cosmogenic nuclide production rates using a continuous geomagnetic model, *Earth Planet. Sci. Lett.* **268** 190–201.

Lippolt, H. J. and Weigel, E., 1988. ^{4}He diffusion in ^{40}Ar-retentive minerals, *Geochim. Cosmochim. Acta* **52** 1449–1458.

Liu, B., Phillips, F. M., Fabryka-Martin, J. T., Fowler, M. M. and Stone, W. D., 1994a. Cosmogenic ^{36}Cl accumulation in unstable landforms: 1. Effects of the thermal neutron distribution, *Water Resour. Res.* **30** 3115–3125.

Liu, B. L., Phillips, F. M., Elmore, D. and Sharma, P., 1994b. Depth dependence of soil carbonate accumulation based on cosmogenic ^{36}Cl dating, *Geology* **22** 1071–1074.

Lucas, L. L. and Unterweger, M. P., 2000. Comprehensive review and critical evaluation of the half-life tritium, *J. Res. Natl. Inst. Stand. Technol.* **105** 541–549.

Mamyrin, B. A., Anufriyev, G. S., Kamensky, I. L. and Tolstikhin, I. N., 1970. Determination of the isotopic composition of helium, *Geochem. Int.* **7** 498–505.

Margerison, H. R., Phillips, F. M., Stuart, F. M. and Sugden, D. E., 2004. Cosmogenic ^{3}He concentrations in ancient flood deposits from the Coombs Hills, Northern Dry Valleys, East Antarctica: interpreting exposure ages and erosion rates, *Earth Planet. Sci. Lett.* **230** 163–175.

Marquette, G. C., Gray, J. T., Gosse, J. C., Courchesne, F., Stockli, L., Macpherson, G. and Finkel, R., 2004. Felsenmeer persistence under non-erosive ice in the Torngat and Kaumajet mountains, Quebec and Labrador, as determined by soil weathering and cosmogenic nuclide exposure dating, *Can. J. Earth Sci.* **41** 19–38.

Martel, D. J., O'Nions, R. K., Hilton, D. R. and Oxburgh, E. R., 1990. The role of element distribution in production and release of radiogenic helium: The Carnmenellis Granite, southwest England, *Chem. Geol.* **88** 207–221.

Marti, K. and Craig, H., 1987. Cosmic-ray-produced neon and helium in the summit lavas of Maui, *Nature* **325** 335–337.

Masarik, J., 2002. Numerical simulation of in situ production of cosmogenic nuclides, *Geochim. Cosmochim. Acta* **66** A491.

Masarik, J. and Beer, J., 1999. Simulation of particle fluxes and cosmogenic nuclide production in the Earth's atmosphere, *J. Geophys. Res.* **104** D 12099–12111.

Masarik, J., Frank, M., Schäfer, J. M. and Wieler, R., 2001. 800 kyr calibration of in-situ cosmogenic nuclide production for geomagnetic field intensity variations, *Geochim. Cosmochim. Acta* **65** 2995–3003.

Masarik, J., Kollar, D. and Vanua, S., 2000. Numerical simulations of in situ production of cosmogenic nuclides: Effects of radiation geometry, *Nucl. Instr. Meth. Phys. Res. B* **172** 786–789.

Masarik, J. and Reedy, R. C., 1994. Effects of bulk composition on nuclide production processes in meteorites, *Geochim. Cosmochim. Acta* **58** 5307–5317.

Masarik, J. and Reedy, R. C., 1995. Terrestrial cosmogenic-nuclide production systematics calculated from numerical simulations, *Earth Planet. Sci. Lett.* **136** 381–395.

Masarik, J. and Reedy, R. C., 1996. Monte Carlo simulation of in-situ produced cosmogenic nuclides, *Radiocarbon* **38** 163–164.

Matmon, A., Crouvi, O., Enzel, Y., Bierman, P., Larsen, J., Porat, N., Amit, R. and Caffee, M., 2003. Complex exposure histories of chert clasts in the late Pleistocene shorelines of Lake Lisan, southern Israel, *Earth Surf. Process. Landforms* **28** 493–506.

Matmon, A., Schwartz, D. P., Haeussler, P. J., Finkel, R., Lienkaemper, J. J., Stenner, H. D. and Dawson, T. E., 2006. Denali fault slip rates and Holocene-late Pleistocene kinematics of central Alaska, *Geology* **34** 645–648.

McDougall, I. and Harrison, T. M., 1999. *Geochronology and Thermochronology by the $^{40}Ar/^{39}Ar$ Method*, Oxford: Oxford University Press, 269 pp.

Merchel, S., Arnold, M., Aumitre, G., Benedetti, L., Bourlès, D. L., Braucher, R., Alfimov, V., Freeman, S. P. H. T., Steier, P. and Wallner, A., 2008a. Towards more precise ^{10}Be and ^{36}Cl data from measurements at the 10^{-14} level: Influence of sample preparation, *Nucl. Instr. Meth. Phys. Res. B* **266** 4921–4926.

Merchel, S., Braucher, R., Benedetti, L., Grauby, O. and Bourlès, D., 2008b. Dating carbonate rocks with in-situ produced cosmogenic Be-10: Why it often fails, *Quat. Geochronol.* **3** 299–307.

Meriaux, A. S., Sieh, K., Finkel, R. C., Rubin, C. M., Taylor, M. H., Meltzner, A. J. and Ryerson, F. J., 2009. Kinematic behavior of southern Alaska constrained by westward decreasing postglacial slip rates on the Denali Fault, Alaska, *J. Geophys. Res. Solid Earth* **114** B03404.

Merrill, R. T., McElhinny, M. W. and McFadden, P. L., 1998. *The Magnetic Field of the Earth*, San Diego: Academic Press, 531 pp.

Michel, R., Leya, I. and Borges, L., 1996. Production of cosmogenic nuclides in meteoroids: accelerator experiments and model calculations to decipher the cosmic ray record in extraterrestrial matter, *Nucl. Inst. Meth. Phys. Res. B* **113** 434–444.

Michel, R. and Neumann, S., 1998. Interpretation of cosmogenic nuclides in meteorites on the basis of accelerator experiments and physical model calculations, *Proc. Indian Acad. Sci. Earth Planet. Sci.* **107** 441–457.

Miller, G. H., Birner, J. P., Lifton, N. A. and Finkel, R. C., 2006. Limited ice-sheet erosion and complex in situ cosmogenic Be-10, Al-26, and C-14 on Baffin Island, Arctic Canada, *Quat. Geochronology* **1** 74–85.

Minasny, B., McBratney, A. B. and Salvador-Blanes, S., 2008. Quantitative models for pedogenesis: A review, *Geoderma* **144** 140–157.

Mitchell, S. G., Matmon, A., Bierman, P. R., Enzel, Y., Caffee, M. and Rizzo, D., 2001. Displacement history of a limestone normal fault scarp, northern Israel, from cosmogenic ^{36}Cl, *J. Geophys. Res.* **106**(B3) 4247–4264.

Mook, W. G. and van der Plicht, J., 1999. Reporting ^{14}C activities and concentrations, *Radiocarbon* **41** 227–239.

Morris, J. D., Leeman, W. P. and Tera, F., 1990. The subducted component in island arc lavas: constraints from Be-isotopes and B-Be systematics, *Nature* **344** 31–36.

Morrison, P. and Pine, J., 1955. Radiogenic origin of the helium isotopes in rock, *Ann. N.Y. Acad. Sci.* **62** 71–92.

Mursula, K., Usoskin, I. G. and Kovaltsov, G. A., 2002. A 22-year cycle in sunspot activity, *Adv. Space Res.* **29** 1979–1984.

Muscheler, R., Joos, F., Beer, J., Müller, S. A., Vonmoos, M. and Snowball, I., 2007. Solar activity during the last 1000 yr inferred from radionuclide records, *Quat. Sci. Rev.* **26** 82–97.

Muscheler, R., Joos, F., Müller, S. A. and Snowball, I., 2005. How unusual is today's solar activity?, *Nature* **436** E3–E4.

Nakamura, T., Uwamino, Y., Ohkubo, T. and Hara, A., 1987. Altitude variation of cosmic ray neutrons, *Health Phys.* **53** 509–517.

Ney, P., 1986. *Gesteinsaufbereitung im Labor*, Stuttgart: Enke, 154 pp.

Nichols, K. K., Bierman, P. R., Eppes, M. C., Caffee, M., Finkel, R. and Larsen, J., 2007. Timing of surficial process changes down a Mojave Desert piedmont, *Quat. Res.* **68** 151–161.

Niedermann, S., 2000. The ^{21}Ne production rate in quartz revisited, *Earth Planet. Sci. Lett.* **183** 361–364.

Niedermann, S., 2002. Cosmic-ray-produced noble gases in terrestrial rocks: dating tools for surface processes, *Rev. Mineral. Geochem.* **47** 731–784.

Niedermann, S., Graf, T., Kim, J. S., Kohl, C. P., Marti, K. and Nishiizumi, K., 1994. Cosmic-ray-produced ^{21}Ne in terrestrial quartz: the neon inventory of Sierra Nevada quartz separates, *Earth Planet. Sci. Lett.* **125** 341–355.

Niedermann, S., Graf, T. and Marti, K., 1993. Mass spectrometric identification of cosmic-ray-produced neon in terrestrial rocks with multiple neon components, *Earth Planet. Sci. Lett.* **118** 65–73.

Niedermann, S., Schaefer, J. M., Wieler, R. and Naumann, R., 2007. The production of cosmogenic ^{38}Ar from calcium in terrestrial pyroxene, *Earth Planet. Sci. Lett.* **257** 596–608.

Niemi, N. A., Oskin, O., Burbank, D., Heimsath, A. M. and Gabet, E. J., 2005. Effects of bedrock landslides on cosmogenically determined erosion rates, *Earth Planet. Sci. Lett.* **237** 480–498.

Nier, A. O., 1947. A mass spectrometer for isotope and gas analysis, *Rev. Sci. Instrum.* **18** 398–411.

Nier, A. O., 1950. A redetermination of the relative abundances of the isotopes of carbon, nitrogen, oxygen, argon and potassium, *Phys. Rev.* **77** 789–793.

Nishiizumi, K., 2004. Preparation of ^{26}Al AMS standards, *Nucl. Instr. Meth. Phys. Res. B* **223** 388–392.

Nishiizumi, K., Caffee, M. W. and DePaolo, D. J., 2000. Preparation of ^{41}Ca AMS standards, *Nucl. Instr. Meth. Phys. Res. B* **172** 399–403.

Nishiizumi, K., Caffee, M. W., Finkel, R. C., Brimhall, G. and Mote, G., 2005. Remnants of a fossil alluvial fan landscape of Miocene age in the Atacama desert of northern Chile using cosmogenic nuclide exposure age dating, *Earth Planet. Sci. Lett.* **237** 499–507.

Nishiizumi, K., Finkel, R. C., Caffee, M. W., Southon, J. R., Kohl, C. P., Arnold, J. R., Olinger, C. T., Poths, J. and Klein, J., 1994. Cosmogenic production of ^{10}Be and ^{26}Al on the surface of the Earth and underground, *Proceedings of the 8th International Conference on Geochronology, Cosmochronology and Isotope Geology*, US Geological Survey Circular 1107 234.

Nishiizumi, K., Finkel, R. C., Klein, J. and Kohl, C. P., 1996. Cosmogenic production of ^{7}Be and ^{10}Be in water targets, *J. Geophys. Res.* **101** 22225–22232.

Nishiizumi, K., Kohl, C. P., Arnold, J. R., Klein, J., Fink, D. and Middleton, R., 1991a. Cosmic ray produced ^{10}Be and ^{26}Al in Antarctic rocks: exposure and erosion history, *Earth Planet. Sci. Lett.* **104** 440–454.

Nishiizumi, K., Imamura, M., Caffee, M. W., Southon, J. R., Finkel, R. C. and McAninch, J., 2007. Absolute calibration of Be-10 AMS standards, *Nucl. Instr. Meth. Phys. Res. B* **258** 403–413.

Nishiizumi, K., Klein, J., Middleton, R. and Craig, H., 1987. *In situ* produced ^{10}Be and ^{26}Al in olivine from Maui, *EOS* **68** 1268.

Nishiizumi, K., Klein, J., Middleton, R. and Craig, H., 1990. Cosmogenic ^{10}Be, ^{26}Al and ^{3}He in olivine from Maui lavas, *Earth Planet. Sci. Lett.* **98** 263–266.

Nishiizumi, K., Kohl, C. P., Arnold, J. R., Dorn, R., Klein, J., Fink, D. and Middleton, R., 1993. Role of in situ cosmogenic nuclides ^{10}Be and ^{26}Al in the study of diverse geomorphic processes, *Earth Surf. Process. Landforms* **18** 407–425.

Nishiizumi, K., Kohl, C. P., Arnold, J. R., Klein, J., Fink, D. and Middleton, R., 1991b. Cosmic ray produced ^{10}Be and ^{26}Al in Antarctic rocks: exposure and erosion history, *Earth Planet. Sci. Lett.* **104** 440–454.

Nishiizumi, K., Lal, D., Klein, J., Middleton, R. and Arnold, J. R., 1986. Production of ^{10}Be and ^{26}Al by cosmic rays in terrestrial quartz in situ and implications for erosion rates, *Nature* **319** 134–135.

Nishiizumi, K., Welten, K. C., Matsumura, H., Caffee, M., Ninomiya, K., Omoto, T., Nakagaki, R., Shima, T., Takahashi, N., Sekimoto, S., Yashima, H., Shibata, S., Bajo, K., Nagao, K., Kinoshita, N., Imamura, M., Sisterson, J. M., Shinohara, A. 2009. Measurements of high-energy neutron cross sections for accurate cosmogenic nuclide production rates, *Geochim. Cosmochim. Acta.* **73** A945.

Nishiizumi, K., Winterer, E. L., Kohl, C. P., Klein, J., Middleton, R., Lal, D. and Arnold, J. R., 1989. Cosmic ray production rates of ^{10}Be and ^{26}Al in quartz from glacially polished rocks, *J. Geophys. Res.* **94** 17907–17915.

Norton, K. P. and Vanacker, V., 2009. Effects of terrain smoothing on topographic shielding correction factors for cosmogenic nuclide-derived estimates of basin-averaged denudation rates, *Earth Surf. Process. Landforms* **34** 145–154.

Ochs, M. and Ivy-Ochs, S., 1997. The chemical behavior of Be, Al, Fe, Ca, and Mg during AMS target preparation from terrestrial silicates modeled with chemical speciation calculations, *Nucl. Instr. Meth. Phys. Res. B* **123** 235–240.

Ohno, M. and Hamano, Y., 1993. Global analysis of the geomagnetic field: Time variation of the dipole moment and the geomagnetic pole in the Holocene, *J. Geomag. Geoeletr.* **45** 1455–1466.

Onuchin, A. A. and Burenina, T. A., 1996. Climatic and geographic patterns in snow density dynamics, *N. Eurasia Arctic Alpine Res.* **28** 99–103.

Owen, L. A., Bright, J., Finkel, R. C., Jaiswal, M. K., Kaufman, D. S., Mahan, S., Radtke, U., Schneider, J. S., Sharp, W., Singhvi, A. K. and Warren, C. N., 2007. Numerical dating of a Late Quaternary spit-shoreline complex at the northern end of Silver Lake playa, Mojave Desert, California: A comparison of the applicability of radiocarbon, luminescence, terrestrial cosmogenic nuclide, electron spin resonance, U-series and amino acid racemization methods, *Quat. Int.* **166**, 87–110.

Ozima, M. and Podosek, F. A., 2001 *Noble Gas Geochemistry*, Cambridge: Cambridge University Press, 300 pp.

Palumbo, L., Benedetti, L., Bourlès, D., Cinque, A. and Finkel, R., 2004. Slip history of the Magnola fault (Apennines, Central Italy) from Cl-36 surface exposure dating: evidence for strong earthquakes over the Holocene, *Earth Planet. Sci. Lett.* **225** 163–176.

Partridge, T. C., Granger, D. E., Caffee, M. W. and Clarke, R. J., 2003. Pliocene hominid remains from Sterkfontein, *Science* **300** 607–612.

Perg, L., Anderson, R. S. and Finkel, R. C., 2001. Use of a new Be-10 and Al-26 inventory method to date marine terraces, Santa Cruz, California, USA, *Geology* **29** 879–882.

Phillips, F. M., Leavy, B. D., Jannik, N. O., Elmore, D. and Kubik, P. W., 1986. The accumulation of cosmogenic chlorine-36 in rocks: a method for surface exposure dating, *Science* **231** 41–43.

Phillips, F. M. and Plummer, M. A., 1996. CHLOE: A program for interpreting in-situ cosmogenic nuclide data for surface exposure dating and erosion studies, *Radiocarbon* **38** 98–99.

Phillips, F. M., Stone, W. D. and Fabryka-Martin, J. T., 2001. An improved approach to calculating low-energy cosmic ray neutron fluxes near the land/atmosphere interface, *Chem. Geol.* **175** 689–701.

Phillips, W. M., Hall, A. M., Mottram, R., Fifield, L. K. and Sugden, D. E., 2006. Cosmogenic Be-10 and Al-26 exposure ages of tors and erratics, Cairngorm Mountains, Scotland: Timescales for the development of a classic landscape of selective linear glacial erosion, *Geomorphol.* **73** 222–245.

Pinti, D., Matsuda, J. and Maruyama, S., 2001. Anomalous xenon in Achean cherts from Pilbara Craton, Western Australia, *Chem. Geol.* **175** 387–395.

Placzek, C., Granger, D. and Caffee, M., 2007. Radiogenic Al-26 chronometry of evaporites, *Geochim. Cosmochim. Acta* **71** A765.

Plug, L. J., Gosse, J., McIntosh, J. J. and Bigley, R., 2007. Attenuation of cosmic ray flux in temperate forest, *J. Geophys. Res. Earth Surf.* **112** F02022.

Porcelli, D., Ballentine, C. J. and Wieler, R., 2002. An overview of noble gas geochemistry and cosmochemistry, *Rev. Mineral. Geochem.* **47** 1–20.

Powell, C. F., Fowler, P. H. and Perkins, D. H., 1959. *The Study of Elementary Particles by the Photographic Method*, London: Pergamon, 669 pp.

Putkonen, J. and Swanson, T., 2003. Accuracy of cosmogenic ages for moraines, *Quat. Res.* **59** 255–261.

Quezada, J., González López, G. A., Dunai, T. J. and Jensen, A., 2007. Edades ^{21}Ne de la terraza costera mas alt del area de Caldera-Bahia Inglesa: Su relacion con el alzamiento litoral pleistoceno del Norte de Chile, *Revista Geol. Chile* **34** 81–96.

Reedy, R. C., Arnold, J. R. and Lal, D., 1983. Cosmic-ray record in solar system matter, *Science* **219** 127–134.

Reid, J. B., Bucklin, E. P., Copenagle, L., Kidder, J., Pack, S. M., Polissar, P. J. and Williams, M. L., 1995. Sliding rocks at the racetrack, Death Valley – What makes them move, *Geology* **23** 819–822.

Reiners, P. W. and Farley, K. A., 1999. Helium diffusion and (U-Th)/He thermo-chronometry of titanite, *Geochim. Cosmochim. Acta* **63** 3845–3859.

Reiners, P. W. and Farley, K. A., 2000. Helium diffusion and (U-Th)/He thermo-chronometry of zircon.

Renne, P. R., Farley, K. A., Becker, T. A. and Sharp, W. D., 2001. Terrestrial cosmogenic argon, *Earth Planet. Sci. Lett.* **188** 435–440.

Repka, J. L., Anderson, R. S. and Finkel, R. C., 1997. Cosmogenic dating of fluvial terraces, Fremont River, Utah, *Earth. Planet. Sci. Lett.* **152** 59–73.

Reynolds, J. H., 1956. High sensitivity mass spectrometer for noble gas analysis, *Rev. Sci. Instrum.* **27** 928–934.

Riebe, C. S., Kirchner, J. W. and Finkel, R. C., 2004. Erosional and climatic effects on long-term chemical weathering rates in granitic landscapes spanning diverse climate regimes, *Earth Planet. Sci. Lett.* **224** 547–562.

Riebe, C. S., Kirchner, J. W. and Granger, D. E., 2001. Quantifying quartz enrichment and its consequences for cosmogenic measurements of erosion rates from alluvial sediment and regolith, *Geomorphol.* **40** 15–19.

Riihimaki, C. and Libarkin, J. C., 2007. Terrestrial cosmogenic nuclides as paleoaltimetric proxies, *Rev. Mineral. Geochem.* **66** 269–278.

Rinterknecht, V. R., Clark, P. U., Raisbeck, G. M., Yiou, F., Bitinas, A., Brook, E. J., Marks, L., Zelcs, V., Lunka, J. P., Pavlovskaya, I. E., Piotrowski, J. A. and Raukas, A., 2006. The last deglaciation of the Southeastern sector of the Scandinavian Ice Sheet, *Science* **311** 1449–1452.

Ross, S. M., 2003. Peirce's criterion for the elimination of suspect experimental data, *J. Eng. Technol.* **20** 38–41.

Rothwell, P., 1958. Cosmic rays in the Earth's magnetic field, *Phil. Mag.* **3** 961–970.

Ruszkiczay-Rüdiger, Z., Dunai, T. J., Bada, G., Fodor, L. and Horváth, E., 2005. Middle to late Pleistocene uplift rate of the Hungarian Mountain Range at the Danube Bend (Pannonian Basin) using in situ produced ^{3}He, *Tectonophys.* **410** 173–187.

Ryerson, F., Tapponnier, P., Finkel, R., Meriaux, A. S., van der Woerd, J., Lasserre, C., Chevalier, M. L., Xu, X., Li, H. and King, G. C. P., 2006. Applications of morphochronology to the active tectonics of Tibet. In: L. Siame, D. L. Bourles and E. T. Brown, (Eds), *In Situ-Produced Cosmogenic Nuclides and Quantification of Geological Surfaces*, Special Paper 415, Boulder: The Geological Society of America, pp. 1–16.

Sarda, P., Staudacher, T. and Allegre, C. J., 1992. Cosmogenic ^{3}He and ^{21}Ne in olivines from Reunion Island: Measurement of erosion rate, *EOS, Supplement* **73** 610.

Sarda, P., Staudacher, T., Allegre, C. J. and Lecomte, A., 1993. Cosmogenic neon and helium at Reunion: measurement of erosion rate, *Earth Planet. Sci. Lett.* **119** 405–417.

Scarsi, P., 2000. Fractional extraction of helium by crushing of olivine and clinopyroxene phenocrysts: Effects on the ^{3}He/^{4}He measured ratio, *Geochim. Cosmochim. Acta* **64** 3751–3762.

Schaefer, J. M., Denton, G., Kaplan, M. R., Putnam, A., Finkel, R. C., Barrell, J. A., Andersen, B. G., Schwartz, R., Macintosh, A., Chinn, T. and Schlüchter, C., 2009. High-frequency Holocene glacier fluctuations in New Zealand differ from the northern signature, *Science* **324** 622–625.

Schäfer, J. M., Faestermann, T., Herzog, G. F., Knie, K., Korschinek, G., Masarik, J., Meier, A., Poutivtsev, M., Rugel, G., Schluchter, C., Serifiddin, F. and Winckler, G., 2006. Terrestrial manganese-53 – A new monitor of Earth surface processes, *Earth Planet. Sci. Lett.* **251** 334–345.

Schäfer, J. M., Ivy-Ochs, S., Wieler, R., Leya, I., Baur, H., Denton, G. H. and Schlüchter, C., 1999. Cosmogenic noble gas studies in the oldest landscape on Earth: surface exposure ages of the Dry Valleys, Antarctica, *Earth Planet. Sci. Lett.* **167** 215–226.

Schaller, M. and Ehlers, T. A., 2006. Limits to quantifying climate driven changes in denudation rates with cosmogenic radionuclides, *Earth Planet. Sci. Lett.* **248** 153–167.

Schaller, M., Ehlers, T. A., Blum, J. D. and Kallenberg, M. A., 2009. Quantifying glacial moraine age, denudation, and soil mixing with cosmogenic nuclide depth profiles, *J. Geophys. Res. Earth Surf.* **114** F01012.

Schaller, M., Hovius, N., Willett, S. D., Ivy-Ochs, S., Synal, H. A. and Chen, M. C., 2005. Fluvial bedrock incision in the active mountain belt of Taiwan from in situ-produced cosmogenic nuclides, *Earth Surf. Process. Landforms* **30** 955–971.

Schaller, M., von Blanckenburg, F., Veldkamp, A., Tebbens, L. A., Hovius, N. and Kubik, P. W., 2002. A 30,000 yr record of erosion rates from cosmogenic Be-10 in Middle European river terraces, *Earth Planet. Sci. Lett.* **204** 307–320.

Schimmelpfennig, I., Benedetti, L., Finkel, R. C., Pik, R., Blard, P. H., Bourlès, D., Burnard, P. G. and Williams, A. J., 2009. Sources of in-situ ^{36}Cl in basaltic rocks. Implications for calibration of production rates, *Quat. Geochronology* doi:10.1016/j.quageo.2009.1004.1003.

Schroeder, P. A., Melear, N. D., Bierman, P., Kashgarian, M. and Caffee, M. W., 2001. Apparent gibbsite growth ages for regolith in the Georgia Piedmont, *Geochim. Cosmochim. Acta* **65** 381–386.

Seidl, M. A., Finkel, R. C., Caffee, M. W., Hudson, G. B. and Dietrich, W. E., 1997. Cosmogenic isotope analyses applied to river longitudinal profile evolution: problems and interpretations, *Earth Surf. Process. Landforms* **22** 195–209.

Serber, R., 1947. Nuclear reactions at high energies, *Phys. Rev.* **72** 1114–1115.

Sharma, P. and Middleton, R., 1989. Radiogenic production of ^{10}Be and ^{26}Al in uranium and thorium ores: Implications for studying terrestrial samples containing low levels of ^{10}Be and ^{26}Al, *Geochim. Cosmochim. Acta* **53** 709–716.

Shim, H., 2002. Corner effect on chloride ion diffusion in rectangular concrete media, *KSCE J. Civil Eng.* **6** 19–24.

Shuster, D. L. and Farley, K. A., 2005. Diffusion kinetics of proton-induced Ne-21, He-3, and He-4 in quartz, *Geochim. Cosmochim. Acta* **69** 2349–2359.

Siame, L. L., Bourles, D. L., Sebier, M., Bellier, O., Castano, J. C., Araujo, M., Perez, M., Raisbeck, G. M. and Yiou, F., 1997. Cosmogenic dating ranging from 20 to 700 ka of a series of alluvial fan surfaces affected by the El Tigre fault, Argentina, *Geology* **25** 975–978.

Simpson, J. A., 1951. Neutrons produced in the atmosphere by the cosmic radiations, *Phys. Rev.* **83** 1175–1188.

Simpson, J. A., 1958. *Cosmic Radiation Neutron Intensity Monitor*, London: Pergamon Press, 351 pp.

Sisterson, J. M., 2005. Cross-section measurements for proton- and neutron-induced reactions needed to understand cosmic-ray interactions on earth and in space. In: R. C. Haight, M. B. Chadwick, T. Kawano and P. Talou, (Eds), *International Conference on Nuclear Data for Science and Technology* **769**, Santa Fe: AIP Conference Procedings, pp. 1596–1599.

Skelton, R. T. and Kavanagh, R. W., 1987. ^{26}Mg(p,n)^{26}Al and ^{23}Na(α,n)^{26}Al reactions, *Phys. Rev. C* **35** 45–54.

Small, E. E., Anderson, R. S. and Hanock, G. S., 1999. Estimates of the rate of regolith production using ^{10}Be and ^{26}Al from alpine hillslope, *Geomorphol.* **27** 131–150.

Smart, D. F., Shea, M. A. and Flückiger, E. O., 2000. Magnetospheric models and trajectory computations, *Space Sci. Rev.* **93** 305–333.

Solanki, S. K., Usoskin, I. G., Kromer, B., Schüssler, M. and Beer, J., 2004. Unusual activity of the Sun during recent decades compared to the previous 11 000 years, *Nature* **431** 1084–1087.

Solanki, S. K., Usoskin, I. G., Kromer, B., Schüssler, M. and Beer, J., 2005. How unusual is today's solar activity? (reply), *Nature* **436** E3–E4.

Srinivasan, B., 1976. Barites: anomalous xenon from spallation and neutron induced reactions, *Earth Planet. Sci. Lett.* **31** 129–141.

Staiger, J., Gosse, J., Toracinta, R., Oglesby, B., Fastook, J. and Johnson, J. V., 2007. Atmospheric scaling of cosmogenic nuclide production: Climate effect, *J. Geophys. Res.* **112** B02205.

Steinhilber, F., Abreu, J. A. and Beer, J., 2008. Solar modulation during the Holocene, *Astrophys. Space Sci. Trans.* **4** 1–6.

Stock, G. S., Anderson, E. C. and Finkel, R. C., 2004. Pace of landscape evolution in the Sierra Nevada, California, revealed by cosmogenic dating of cave sediments, *Geology* **32** 193–196.

Stone, J., Allan, G. L., Fifield, L. K., Evans, J. M. and Chivas, A. R., 1994. Limestone erosion measurements with cosmogenic chlorine-36 in calcite – preliminary results from Australia, *Nucl. Instr. Meth. B* **92** 311–316.

Stone, J., Evans, J., Fifield, K., Cresswell, R. and Allan, G., 1996a. Cosmogenic chlorine-36 production rates from calcium and potassium, *Radiocarbon* **38** 170–171.

Stone, J., Lambeck, K., Fifield, L. K., Evans, J. M. and Cresswell, R. G., 1996b. A late glacial age for the Main Rock Platform, western Scotland, *Geology* **24** 707–710.

Stone, J. O., 1998. A rapid fusion method for separation of beryllium-10 from soils and silicates, *Geochim. Cosmochim. Acta* **62** 555–561.

Stone, J. O., 2000. Air pressure and cosmogenic isotope production, *J. Geophys. Res.* **105** 23753–23759.

Stone, J. O., 2005. *Terrestrial Chlorine-36 Production from Spallation of Iron*, 10th AMS Conference, Berkeley.

Stone, J. O., Allan, G. L., Fifield, L. K. and Cresswell, R. G., 1996c. Cosmogenic chlorine-36 from calcium spallation, *Geochim. Cosmochim. Acta* **60** 679–692.

Stone, J. O., Evans, J. M., Fifield, L. K., Cresswell, R. G. and Allan, G. L., 1996d. Cosmogenic Cl-36 production rates from calcium and potassium, *Radiocarbon* **38** 170–170.

Stone, J. O., Evans, N. J., Fifield, L. K., Allan, G. L. and Cresswell, R. G., 1998. Cosmogenic chlorine-36 production in calcite by muons, *Geochim. Cosmochim. Acta* **62** 433–454.

Streckeisen, A., 1973. To each plutonic rock its proper name, *Earth Sci. Rev.* **12** 1–33.

Stuart, F. M., Lass-Evans, S., Fitton, J. G. and Ellam, R. M., 2003. High He-3/He-4 ratios in picritic basalts from Baffin Island and the role of a mixed reservoir in mantle plumes, *Nature* **424** 57–59.

Stuiver, M. and Polach, H., 1977. Reporting of ^{14}C data, *Radiocarbon* **19** 355–363.

Suter, M., 2004. 25 years of AMS: A review of recent developments, *Nucl. Instr. Meth. Phys. Res. B* **223–224** 139–148.

Swanson, T. W. and Caffee, M., 2001. Determination of ^{36}Cl production rates derived from the well-dated deglaciation surfaces of Whidbey and Fidalgo Islands, Washington, *Quat. Research* **56** 366–382.

Taggart, A. F., 1945. *Handbook of Mineral Dressing, Ores and Industrial Minerals*, London: Chapman & Hall, 1915 pp.

Taylor, J. R., 1997. *An Introduction to Error Analysis*, Mill Valley, CA: University Science Books, 327 pp.

Taylor, R. E. and Berger, R., 1967. Radiocarbon content of marine shells from Pacific Coast of Central and South America, *Science* **158** 1180–1182.

Teucher, M., 1952. Die Absorption der Nukleonenkomponente der kosmischen Strahlung in Luft zwischen Seehöhe und 4000 m, *Z. Naturforschg.* **7a** 61–63.

Trull, T. W., Kurz, M. D. and Jenkins, W. J., 1991. Diffusion of cosmogenic ^{3}He in olivine and quartz: implications for surface exposure dating, *Earth Planet. Sci. Lett.* **103** 241–256.

Usoskin, I. G., Alanko-Huotari, K., Kovaltsov, G. A. and Mursula, K., 2005. Heliospheric modulation of cosmic rays: Monthly reconstruction for 1951–2004, *J. Geophys. Res.* **110** A12108.

Valet, J. P., Herrero-Bervera, E., LeMouel, J. L. and Plenier, G., 2008. Secular variations of the geomagnetic dipole during the past 2000 years, *Geochem. Geophys. Geosyst.* **9** Q01008.

Valet, J. P., Meynadier, L. and Guyodo, Y., 2005. Geomagnetic dipole strength and reversal rate over the past two million years, *Nature* **435** 802–805.

Van der Plicht, J. and Hogg, A., 2006. A note on reporting radiocarbon, *Quat. Geochrono.* **1** 237–240.

Van der Wateren, F. M. and Dunai, T. J., 2001. Late Neogene passive margin denudation history: cosmogenic isotope measurements from the Central Namib desert, *Global Planet. Change* **30** 271–307.

Van der Wateren, F. M., Dunai, T. J., Van Balen, R. T., Klas, W., Verbers, A. L. L. M., Passchier, S. and Herpers, U., 1999. Contrasting neogene denudation histories of different structural regions in the Transantarctic Mountains rift flank constrained by cosmogenic isotope measurements, *Global Planet. Change* **23** 145–172.

Vanacker, V., von Blanckenburg, F., Govers, G., Molina, A., Poesen, J., Deckers, J. and Kubik, P. W., 2007. Restoring dense vegetation can slow mountain erosion to near natural benchmark levels, *Geology* **35** 303–306.

Vega-Carillo, H. R., Manzanares-Acuna, E., Hernandes-Davila, V. M. and Sanchez, G. A. M., 2005. Response matrix of a multisphere neutron spectrometer with an He-3 proportional counter, *Revista Mex. Fisica* **51** 47–52.

Vermeesch, P., 2007. CosmoCalc: An Excel add-in for cosmogenic nuclide calculations, *Geochem. Geophys. Geosyst.* **8** Q08003.

Vermeesch, P., Heber, V., Strasky, S., Kober, F., Schaefer, J. M., Baur, H., Schlüchter, C. and Wieler, R., 2008. Cosmogenic ^{3}He and ^{21}Ne measured in artificial quartz targets after one year of exposure in the Swiss Alps, *Geophys. Res. Abstr.* **10** 1607–7962/gra/EGU2008-A-08431.

Vidyadhar, A., Rao, K. H. and Forssberg, K. S. E., 2002. Separation of feldspar from quartz: Mechanism of mixed cationic/anionic collector adsorption on minerals and flotation selectivity, *Minerals & Metallurgical Processing* **19** 128–136.

Viles, H. A. and Bourke, M. C., 2007. *A Photographic Atlas of Rock Breakdown Features in Geomorphic Environments*, Tucson: Planetary Science Institute.

Villoresi, G., Iucci, N., Re, F., Signoretti, F., Zangrilli, N., Cecchini, S., Parisi, M., Signorini, C., Tyasto, M. I., Danilova, O. A. and Ptitsyna, N. G., 1997. Latitude survey of cosmic ray nucleogenic component during 1996–1997 from Italy to Antarctica. In: S. Aiello, N. Iucci, G. Sironi, A. Treves and U. Villante, (Eds), *8th CIFCO Conference, Cosmic Ray Physics in the Year 2000, Conference Proceedings* **58**, Como: The Italian Physics Society, pp. 323–326.

von Blanckenburg, F., 2005. The control mechanisms of erosion and weathering at basin scale from cosmogenic nuclides in river sediment, *Earth Planet. Sci. Lett.* **237** 462–479.

von Blanckenburg, F., Belshaw, N. S. and O'Nions, R. K., 1996. Separation of ^{9}Be and cosmogenic ^{10}Be from environmental materials and SIMS isotope dilution analysis, *Chem. Geol.* **129** 93–99.

Vries, H. D., 1958. Atomic bomb effect variation of radiocarbon in plants, shells, and snails in the past 4 years, *Science* **128** 250–251.

Wells, S. G., McFadden, L. D., Poths, J. and Olinger, C. T., 1995. Cosmogenic [3]He surface exposure dating of stone pavements, *Geology* **23** 613–616.

Wernicke, R. and Lippolt, H. J., 1994. [4]He age discordance and release behavior of a double shell botryoidal hematite from the Schwarzwald, Germany, *Geochim. Cosmochim. Acta* **58** 421–429.

Wiedenbeck, M. E., Davis, A. J., Leske, R. A., Binns, W. R., Cohen, C. M. S., Cummings, A. C., de Nolfo, G., Israel, M. H., Labrador, A. W., Mewaldt, R. A., Scott, L. M., Stone, E. C. and von Rosenvinge, T. T., 2005. The level of solar modulation of galactic cosmic rays from 1997 to 2005 as derived from ACE measurements of elemental energy spectra, *Proceedings of the 29th International Cosmic Ray Conference, Prune, India, 03–10 August, 2005* **2** Mumbai: Tata Institue of Fudamantal Research, pp. 277–280.

Wieler, R., 2002. Cosmic-ray-produced noble gases in meteorites, *Reviews in Mineralogy and Geochemistry* **47** 125–170.

Wilkinson, B. H. and McElroy, B. J., 2007. The impact of humans on continental erosion and sedimentation, *Geol. Soc. Am. Bull.* **119** 140–156.

Wilkinson, M. T. and Humphreys, G. S., 2005. Exploring pedogenesis via nuclide-based soil production rates and OSL-based bioturbation rates, *Austr. J. Soil Res.* **43** 767–779.

Williams, A. J., Stuart, F. M., Day, S. J. and Phillips, W. M., 2005. Using pyroxene microphenocrysts to determine cosmogenic He-3 concentrations in old volcanic rocks: an example of landscape development in central Gran Canaria, *Quat. Sci. Rev.* **24** 211–222.

Wolokowinsky, F. L. and Granger, D. E., 2004. Early Pleistocene incision of the San Juan River, Utah, dated with [26]Al and [10]Be, *Geology* **32** 749–752.

Yang, S., Odah, H. and Shaw, J., 2000. Variations in the geomagnetic dipole moment of the last 12000 years, *Geophys. J. Int.* **140** 158–162.

Yokoyama, Y., Reyss, J.-L. and Guichard, F., 1977. Production of radionuclides by cosmic rays at mountain altitudes, *Earth Planet. Sci. Lett.* **36** 44–50.

Youngson, J., Bennet, E., Jackson, J., Norris, R., Raisbeck, G. M. and Yiou, F., 2005. 'Sarsen Stones' at German Hill, Central Otago, New Zealand, and their potential for in situ cosmogenic isotope dating of landscape evolution, *J. Geol.* **113** 341–354.

Ziegler, J. F., Biersack, J. P. and Ziegler, M. D., 2008. *SRIM Stopping Ranges of Ions in Matter*, Morrisville: Lulu Press, 398 pp.

Zreda, M., Desilets, D., Ferré, T. P. A. and Scott, R. L., 2008. Measuring soil moisture content non-invasively at intermediate spatial scale using cosmic-ray neutrons, *Geophys. Res. Lett.* **35** L21402.

Zreda, M. and Noller, J., 1998. Ages of prehistoric earthquakes revealed by cosmogenic chlorine-36 in a bedrock fault scarp at Hebgen Lake, *Science* **292** 1097–1099.

Zreda, M. G., Phillips, F. M., Elmore, D., Kubik, P. W., Sharma, P. and Dorn, R. I., 1991. Cosmogenic [36]Cl production in terrestrial rocks, *Earth Planet. Sci. Lett.* **105** 94–105.

Index

Printed in the United States
By Bookmasters